PROGRAMMING LANGUAGES:
AN ACTIVE LEARNING APPROACH

PROGRAMMING LANGUAGES: AN ACTIVE LEARNING APPROACH

Kent D. Lee
Luther College

 Springer

Kent D. Lee
Luther College
Department of Computer Science
700 College Drive
Decorah, Iowa 52101-1045
leekentd@luther.edu

ISBN: 978-1-4419-4636-2   e-ISBN: 978-0-387-79422-8
DOI: 10.1007/978-0-387-79421-1

Printed on acid-free paper

9 8 7 6 5 4 3 2 1

springer.com

To Denise.

To Denise.

Preface

Computer Science has matured into a very large discipline. You will not be taught everything you will need in your career while you are a student. The goal of a Computer Science education is to prepare you for a life of learning. The creativity encouraged by a lifetime of learning makes Computer Science one of the most exciting fields today according to Money Magazine's Best Jobs in America in 2006.

The words *Computer Science* don't really reflect what most computer programmers do. The field really should be named something like Computer Program Engineering. It is more of an engineering field than a science. That's why programmers are often called Software Engineers. There *is* science in Computer Science generally relating to the gathering of empirical evidence pertaining to performance of algorithms and hardware. There are also theoretical aspects of the discipline, some of which this text will explore. However, the bulk of what computer scientists do is related to the creative process of building programs. As computer scientists we know some things about how to write good programs, but it is still a very creative process.

Given that programming is such a creative process, it is imperative that we be able to *predict* what our programs will do. To predict what a program will do, you must understand how the program works. The programs of a language execute according to a model of computation. A model may be implemented in many different ways depending on the targeted hardware architecture. However, it is not necessary to understand all the different architectures out there to understand the model of computation used by a language.

For several years in the late 1980's and 1990's I worked at IBM on the operating system for the AS/400 (what is now called System i and the i5/OS). My understanding of compilers and language implementation helped me make better decisions about how to write code and several times helped me find problems in code I was testing. An understanding of the models of computation used by the languages I programmed in aided me in predicting what my programs would do. After completing a course in programming languages, you should understand some of the basics of language implementation. This book is not intended to be a complete text on compiler or interpreter implementation, but there are aspects of language implementation that are included in it. We are better users of tools when we understand how the tools we use work.

I hope you enjoy learning from this text and the course you are about to take. The text is meant to be used interactively. You should read a section and as you read it, do the practice exercises listed in the gray boxes. Each of the exercises are meant to give you a goal in reading a section of the text.

The text has a website where code and other support files may be downloaded. See http://www.cs.luther.edu/~leekent/ProgrammingLanguages for all support files.

For Teachers

This book was written to fulfill two goals. The first is to introduce students to three programming paradigms: object-oriented/imperative, functional, and logic programming. To be ready for the content of this book students should have some background in an imperative language, probably an object-oriented language like Python or Java. They should have had an introductory course and a course in Data Structures as a minimum. While the prepared student will have written several programs, some of them fairly complex, most probably still struggle with predicting exactly what their program will do. It is assumed that ideas like polymorphism, recursion, and logical implication are relatively new to the student reading this book. The functional and logic paradigms, while not the mainstream, have their place and have been successfully used in interesting applications.

The Object-Oriented languages presented in this book are C++ and Ruby. Teachers may choose between the chapter on Ruby or the chapter on C++, or may assign both. It might be useful to read both chapters if you wish to compare and contrast a statically typed, compiled language to a dynamically typed, interpreted language. The same project is presented in both chapters with the C++ chapter requiring a little more explanation in terms of the compiler and organization of the code. Either language is interesting to choose from and the chapters do cross-reference each other to compare and contrast the two styles of programming so if you only have time to cover one or the other, that is possible too.

C++ has many nuances that are worthy of several chapters in a programming languages book. Notably the pass by value and pass by reference mechanisms in C++ create considerable complexity in the language. Polymorphism is another interesting aspect of Object-Oriented languages that is studied in this text.

Ruby is relatively new to the programming language arena, but is widely accepted and is a large language when compared to Python and Java. In addition, its object-centered approach is very similar to Smalltalk. Ruby is also interesting due to the recent development of "Ruby on Rails" as a code generation tool.

The text uses Standard ML as the functional language. ML has a polymorphic type inference system to statically type programs of the language. In addition, the type inference system of ML is formally proven sound and complete. This has some implications in writing programs. While ML's cryptic compiler error messages are sometimes hard to understand at first, once a program compiles it will often work correctly the first time. That's an amazing statement to make if your past experience is in a dynamically typed language like Lisp, Scheme, Ruby, or Python.

The logic language is Prolog. While Prolog is an Artificial Intelligence language, it originated as a meta-language for expressing other languages. The text concentrates on using Prolog to implement other languages. Students learn about logical implication and how a problem they are familiar with can be re-expressed in a different paradigm.

The second goal of the text is to be interactive. This book is intended to be used in and outside of class. It is my experience that we almost all learn more by doing than by seeing. To that end, the text encourages teachers to actively teach. Each chapter follows a pattern of presenting a topic followed by a practice exercise or exercises that encourage students to try what they have just read. These exercises can be used in class to help students check their understanding of a topic. Teachers are encouraged to take the time to present a topic and then allow students time to practice with the concept just presented. In this way the text becomes a lecture resource. Students get two things out of this. It forces them to be interactively engaged in the lectures, not just passive observers. It also gives them immediate feedback on key concepts to help them determine if they understand the material or not. This encourages them to ask questions when they have difficulty with an exercise. Tell students to bring the book to class along with a pencil and paper. The practice exercises are easily identified. Look for the light gray practice problem boxes.

The book presents several projects to reinforce topics outside the classroom. Each chapter of the text suggests several non-trivial programming projects that accompany the paradigm being covered to drive home the concepts covered in that chapter. The projects described in this text have been tested in practice and documentation and solutions are available upon request.

Finally, it is expected that while teaching a class using this text, lab time will be liberally sprinkled throughout the course as the instructor sees fit. Reinforcing lectures with experience makes students appreciate the difficulty of learning new paradigms while making them stronger programmers, too.

Supplementary materials including sample lecture notes, lecture slides, answers to exercises, and programming assignment solutions are available to instructors upon request.

Acknowledgments

I have been fortunate to have good teachers throughout high school, college, and graduate school. Good teachers are a valuable commodity and we need more of them. Ken Slonneger was my advisor in graduate school and this book came into being because of him. He inspired me to write a text that supports the same teaching style he uses in his classroom. Encouraging students to interact during lecture by giving them short problems to solve that reflect the material just covered is a very effective way to teach. It makes the classroom experience active and energizes the students. Ken graciously let me use his lecture notes on Programming Languages when I started this book and some of the examples in this text come from those notes. He also provided me with feedback on this text and I appreciate all that he did. Thank you very much, Ken!

Other great teachers come to mind as well including Dennis Tack who taught me the beauty of a good proof, Karen Nance who taught me to write, Alan Macdonald who introduced me to programming languages as a field of study, Walt Will who taught me how to write my first assembler, and last but not least Steve Hubbard who still inspires me with his ability to teach complex algorithms and advanced data structures to Computer Science majors at Luther College! Thanks to you all.

Contents

Chapter 1

Introduction

The intent of this text is to introduce you to two new programming paradigms that you probably haven't used before. As you learn to program in these new paradigms you will begin to understand that there are different ways of thinking about problem solving. Each paradigm is useful in some contexts. This book is not meant to be a survey of lots of different languages. Rather, its purpose is to introduce you to the three styles of programming languages. These styles are:

- Imperative/Object-Oriented Programming with languages like Java, C++, Ruby, Pascal, Basic, and other languages you probably have used before.
- Functional Programming with languages like ML, Haskell, Lisp, Scheme, and others.
- Logic Programming with Prolog.

This book includes examples from C++, Ruby, ML, Pascal, and Prolog while touching on other languages as well. The book provides an introduction to programming in C++, Ruby, ML, and Prolog. Each of these languages deserves at least a text of their own. In fact, C++ [11], Ruby[34], Standard ML [36], and Prolog [7] each do have books written about them, many more than are cited here. The goal of the text is to help you understand how to use the paradigms and models of computation these languages represent to solve problems. You should learn when these languages are good choices and perhaps when they aren't good choices. You should also learn techniques for problem solving that you may be able to apply in the future. You might be surprised by what you can do with very little code given the right language.

To begin you should know something about the history of computing, particularly as it applies to the models of computation that have been used in implementing many of the programming languages we use today. All of what we know in Computer Science is built on the shoulders of those who came before us. To understand where we are, we really should know something about where we came from in terms of Computer Science. Many great people have been involved in the development of programming languages and to learn even a little about who these people are is really fascinating and worthy of an entire textbook in itself.

K.D. Lee, *Programming Languages*, DOI: 10.1007/978-0-387-79421-1_1,
© Springer Science+Business Media, LLC 2008

1.1 Historical Perspective

Much of what we attribute to Computer Science actually came from mathematicians. After all, mathematicians are really programmers that have written their programs using mathematical notation. Sophus Lie found ways of solving Ordinary Differential Equations by exploiting properties of symmetry within the equations[12]. [*Sophus was my great-great-grandfather's first cousin. When Johann Hermann Sophian Lie, my great-great-grandfather, immigrated from Norway to the United States the spelling was changed from Lie to Lee but is pronounced the same either way.*] Sophus didn't have a computer available to him as a tool to solve problems. He lived from 1842-1899. While the techniques he discovered were hard for people to learn and use at the time, today computer programs capable of symbolic manipulation use his techniques to solve these and other equally complicated problems. Sophus discovered a set of groups in Abstract Algebra which have been named Lie Groups. One such group, the $E8$ group was too complicated to map in Lie's time. In fact, it wasn't until 2007 that the structure of the $E8$ group could be mapped because the solution produced sixty times more data than the human genome project[10].

As mathematicians' problem solving techniques became more sophisticated and the problems they were solving became more complex, they were interested in finding automated ways of solving these problems. Charles Babbage (1791-1871) saw the need for a computer to do calculations that were too error-prone for humans to perform. He designed a *difference engine* to compute mathematical tables when he found that human *computers* weren't very accurate[38]. However, his computer was mechanical and couldn't be built using engineering techniques known at that time. In fact it wasn't completed until 1990, but it worked just as he said it would over a hundred years earlier.

Fig. 1.1: John Backus[2]

Charles Babbage's difference engine was an early attempt at automating a solution to a problem. In the 1940's people like Atanasoff and Eckert and Mauchly[39] were interested in automated computing machines. With the discovery of the transistor they were finally able to design the precursors to today's modern computers. John von Neumann made one of the great contributions to the architecture of modern computers when, in 1944, he wrote a memo suggesting how to encode programs as sequences of numbers resulting in a stored-program computer. Alan Turing followed in 1946 with a paper describing a complete design for such a computer. To this day the computers we use are stored-program computers. The architecture is called the von Neumann architecture because of John von Neumann's contributions.

In the early days of Computer Science, many programmers were interested in writing tools that made it easier to program computers. Much of the programming was based on the concept of a stored-program computer and many early programming languages were extensions of this model of computation. In the stored-program model the program and data are stored in memory. The program manipulates data based on some input. It then produces output.

Around 1958, Algol was created and the second revision of this language, called Algol 60, was the first modern, structured, imperative programming language. While the language was designed by a committee, a large part of the success of the project is due to the contributions of John Backus pictured in figure 1.1. He described the structure of the Algol language using a mathematical notation that would later be called Backus-Naur Format or BNF. Very little has changed with the underlying computer architecture over the years. Of course, there have been many changes in the size, speed, and cost of computers! In addition, the languages we use have become even more structured over the years. But, the principles that Algol 60 introduced are still in use today.

Recalling that most early computer scientists were mathematicians, it shouldn't be surprising to learn that there were others that approached the problem of programming from a drastically different angle. Much of the initial interest in computers was spurred by the invention of the stored-program computer and many of the early languages reflected this excitement. However, there was another approach developing at the same time. Alonzo Church was developing a language called the lambda calculus, usually written as the λ-calculus.

Ideas from the λ-calculus led to the development of Lisp by John McCarthy, pictured in figure 1.2. The λ-calculus and Lisp were not designed based on the principle of the stored-program computer. In contrast to Algol 60, the focus of these languages was on functions and what could be computed using functions. The goal of this work was to arrive at a language mathematicians

Fig. 1.2: John McCarthy[22]

and computer scientists could use to formally describe the calculations they previously had described using informal language. With the λ-calculus and Lisp you could describe all computable functions. Lisp was developed around 1958, the same time that Algol 60 was being developed.

Of course, with many of the early computer scientists being mathematicians, it is reasonable to suspect that many of them were trying to solve problems involving logic. Languages for logic were well developed in the early twentieth century. Many problems were being expressed in terms of propositional logic and first order predicate calculus. It was natural to look for a way to use computers to solve at least some of these problems.

The foundations of logic are so historic they stretch back to ancient Greece, China, and India. The logic we use today in Western culture originated with Aristotle in Greece. Prolog is a programming language that grew out of the desire to solve problems using logic.

1.2 Models of Computation

The earliest computers were made possible by the concept of a stored program computer. While some controversy exists about this, John von Neumann is generally given credit for coming up with the concept of storing a program as a string of 0's and 1's in memory along with the data used by the program. The von Neumann architecture had very little structure to it. It consisted of several registers and memory. The Program Counter (PC) register kept track of the next instruction to execute. There were other registers that could hold a value or point to other values stored in memory. This model of computation was useful when programs were small and weren't organized using top-down design. Functions and procedures impose more structure on programs than the von Neumann model. In fact, in the early days a function was often called a sub-routine instead of a function. The von Neumann model of computation is characterized by a simple programming language called RAM[29] which stands for Random Access Machine. This language has been slightly extended to provide the language called EWE presented in the next chapter.

The Imperative Model

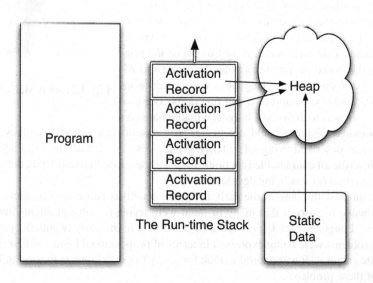

Fig. 1.3: Conceptual View of the Imperative Model

With the advent of Algol 60 some structure was added to the von Neumann model, leading to the structured, imperative model of computing. In the imperative model, memory is divided into regions which hold the program and the data. The data area is further subdivided into the static or global data area, the run-time stack, and the heap pictured in figure 1.3.

When a program executes it uses a special register called the stack pointer (SP) to point to the top activation record on the run-time stack. The run-time stack contains one activation record for each function or procedure invocation that is currently unfinished in the program. The top activation record corresponds to the current function invocation. When a function call is made an activation record is pushed onto the run-time stack. When a function returns, the activation record is popped by decrementing the stack pointer to point to the previous activation record.

An activation record contains information about the currently executing function. The local variables of the function are stored there. The program counter's value before the function call was made is stored there. Other state information may also be stored there depending on the language and the details of the underlying von Neumann architecture. For instance, parameters passed to the function may also be stored there. The return address may also be stored in the activation record.

Static or global data sometimes exists and sometimes does not depending on the language. Where global data is stored depends on the implementation of the compiler or interpreter. It might be part of the program code in some instances. In any case, this area is where constants and global variables are stored. Global variables are those variables that are available to all functions and not just the current function.

The heap is an area for dynamic memory allocation. The word dynamic means that it happens while the program is running. All data that is created at run-time is located in the heap. The data in the heap has no names associated with the values stored there. Instead, named variables called pointers or references point to the data in the heap. In addition, data in the heap may contain pointers that point to other data in the heap.

The primary goal of the imperative model is to get data as input, transform it via updates to memory, and then produce output based on this imperatively changed data. This model of computation parallels the underlying von Neumann architecture. The imperative model of computation can be used by a programming language to provide a structured way of writing programs. Some variation of this model is used by languages like Algol 60, C++, C, Java, VB.net, Python, and many other languages.

☞ Practice 1.1

Find the answers to the following questions.

1. What are the three divisions of data memory called?
2. When does an item in the heap get created?
3. What goes in an activation record?
4. When is an activation record created?
5. When is an activation record deleted?
6. What is the primary goal of imperative, object-oriented programming?

The Functional Model

In the functional model the goal of a program is slightly different. This slight change in the way the model works has a big influence on how you program. In the functional model of computation the focus is on function calls. Functions and parameter passing are the primary means of accomplishing data transformation.

Data is not changed in the functional model. Instead, new values are constructed from old values. A pure functional model wouldn't allow any updates to existing values. However, most functional languages allow limited updates to memory.

The conceptual view presented in figure 1.3 is similar to the view in the functional world. However, the difference between program and data is eliminated. A function is data like any other data element. Integers and functions are both first-class citizens of the functional world.

The static data area may be present, but takes on a minor role in the functional model. The run-time stack becomes more important because most work is accomplished by calling functions. Functional languages are much more careful about how they allow programmers to access the heap and as a result, you really aren't aware of the heap when programming in a functional language. Data is certainly dynamically allocated, but once data is created on the heap it cannot be modified in a pure functional model. Impure models might allow some modification of storage but this is the influence of imperative languages creeping into the functional model as a way to deal with performance issues. The result is that you spend less time thinking about the underlying architecture when programming in a functional language.

☞ Practice 1.2

Answer the following questions.

1. What are some examples of functional languages?
2. What is the primary difference between the functional and imperative models?
3. Immutable data is data that cannot be changed once created. The presence of immutable data simplifies the conceptual model of programming. Does the imperative or functional model emphasize immutable data?

The Logic Model

The logic model of computation, pictured in figure 1.4, is quite different from either the imperative or functional model. In the logic model the programmer doesn't actually write a program at all. Instead, the programmer provides a database of facts or rules. From this database, a single program tries to answer questions with a yes or no answer. In the case of Prolog, the program acts in a predictable manner allowing the programmer to provide the facts in an order that determines how the program will work. The actual implementation of this conceptual view is accomplished by a

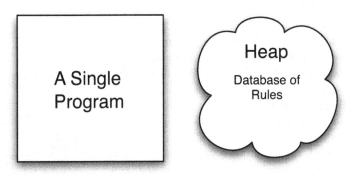

Fig. 1.4: Conceptual View of the Logic Model of Computation

virtual machine, a technique for implementing languages that is covered later in this book.

There is still the concept of a heap in Prolog. One can assert new rules and retract rules as the program executes. To dynamically add rules or retract them there must be an underlying heap. In fact, the run-time stack is there too. However, the run-time stack and heap are so hidden in this view of the world that it is debatable whether they should appear in the conceptual model at all.

☞ Practice 1.3

Answer these questions on what you just read.

1. How many programs can you write in a logic programming language like Prolog?
2. What does the programmer do when writing in Prolog?

1.3 The Origins of a Few Programming Languages

This book explores language implementation using several small languages and exercises that illustrate each of these models of computation. In addition, exercises within the text will require implementation in four different languages: C++, Ruby, Standard ML, and Prolog. But where did these languages come from and why are we interested in learning how to use them?

A Brief History of C++

C++ was designed by Bjarne Stroustrup, pictured in figure 1.5, between 1980 and 1985 while working at Bell Labs. C++ was designed as a superset of C which was an immensely popular language in the seventies and eighties and still is today. In C, the ++ operator is called the increment operator. It adds one to the variable that precedes it. C++ was the next increment after C.

Fig. 1.5: Bjarne Stroustrup[32]

In 1972, the Unix operating system was written in C, which was the reason the language was created. Ken Thompson was working on the design of Unix with Dennis Ritchie. It was their project that encouraged Ritchie to create the C language. C was more structured than the assembly language most operating systems were written in at the time and it was portable and could be compiled to efficient machine code. Thompson and Ritchie wanted an operating system that was portable, small, and well organized.

While C was efficient, there were other languages that had either been developed or were being developed that encouraged a more structured approach to programming. For several years there had been ideas around about how to write code in Object-Oriented form. Simula, created by Ole-Johan Dahl and Kristen Nygaard around 1967, was an early example of a language that supported Object-Oriented design and Modula-2, created by Niklaus Wirth around 1978, was also taking advantage of these ideas. Smalltalk, an interpreted language, was object-oriented and was also developed in the mid 1970's and released in 1980.

Around 1980, Bjarne Stroustrup began working on the design of C++ as a language that would allow C programmers to keep their old code while allowing new code to be written using these Object-Oriented concepts. In 1983 he named this new language C++ and with much anticipation, in 1985 the language was released. About the same time Dr. Stroustrup released a book called *The C++ Programming Language* which described the language. The language was still evolving for a few years. For instance, templates, an important part of C++ were first described by Stroustrup in 1988[31] and it wasn't until 1998 that it was standardized as ANSI C++. Today an ANSI committee oversees the continued development of C++ although changes to the standard have slowed in recent years.

A Brief History of Ruby

Ruby is a relatively new addition as a programming language. Yukihiro Matsumoto (Matz for short) created Ruby in 1993 and released it to the public in 1995. In recent years it has gained wide acceptance. Ruby is an object-oriented scripting language.

Scripting languages are languages that are interpreted and allow the programmer to quickly build a program and test it as it is written. Prototyping, testing, and revising the prototype is a very effective way to program. People like Matz have seen the benefits of this style of programming. Ruby is based on several other interpreted languages like Smalltalk, Python, Perl, and Lisp that are also very popular languages. As Matz was developing Ruby he "wanted a scripting language that was more powerful than Perl, and more object-oriented than Python."[28]

In Ruby, all data are called objects. Every function that is written is a method that operates on an object. The syntax of the language is very large. In contrast, languages like Java are quite small in size. This shouldn't be confused with the availability of classes written in Ruby and Java. Java and Ruby both have large libraries of available classes. Unlike Java, the Ruby language itself is also large. Matz created a large language on purpose because one of his goals was to relieve the programmer from menial tasks in programming.

Ruby is an interesting language to study because it has a large following in Japan and the rest of the world. It is also gaining support among those people developing database applications (which is how a good deal of recent programs store data). The Ruby framework for developing database applications is often referred to as "Ruby on Rails" and again asserts the desire for programmers to be relieved of menial tasks, in this case referring to writing code to interact with a database.

A Brief History of Standard ML

Fig. 1.6: Robin Milner[23]

Standard ML originated in 1986, but was the follow-on of ML which originated in 1973[24]. Like many other languages, ML was implemented for a specific purpose. The ML stands for Meta Language. Meta means above or about. So a metalanguage is a language about language. In other words, a language used to describe a language. ML was originally designed for a theorem proving system. The theorem prover was called LCF, which stands for Logic for Computable Functions. The LCF theorem prover was developed to check proofs constructed in a particular type of logic first proposed by Dana Scott in 1969 and now called Scott Logic. Robin Milner, pictured in figure 1.6, was the principal designer of the LCF system. Milner designed the first version of LCF while at Stanford University. In 1973, Milner moved to Edinburgh University and hired Lockwood Morris and Malcolm Newey, followed by Michael Gordon and Christopher Wadsworth, as research associates to help him build a new and better version called Edinburgh LCF[13].

For the Edinburgh version of LCF, Dr. Milner and his associates created the ML programming language to allow proof commands in the new LCF system to be extended and customized. ML was just one part of the LCF system. However, it quickly became clear that ML could be useful as a general purpose programming language. In 1990 Milner, together with Mads Tofte and Robert Harper, published the first complete formal definition of the language; joined by David MacQueen, they revised this standard to produce the Standard ML that exists today[24].

ML was influenced by Lisp, Algol, and the Pascal programming languages. In fact, ML was originally implemented in Lisp. There are now two main versions of ML: Moscow ML and Standard ML. Today, ML's main use is in academia in the research of programming languages. But, it has been used successfully in several other types of applications including the implementation of the TCP/IP protocol stack and a web server as part of the Fox Project. A goal of the Fox Project is the development of system software using advanced programming languages[14].

An important facet of ML is the strong type checking provided by the language. The type inference system, commonly called Hindley-Milner type inference, statically checks the types of all expressions in the language. In addition, the type checking system is polymorphic, meaning that it handles types that may contain type variables. The polymorphic type checker is sound. It will never say a program is typed correctly when it is not. Interestingly, the type checker has also been proven complete, which means that all correctly typed programs will indeed pass the type checker. No correctly typed program will be rejected by the type checker. We expect soundness out of type checkers but completeness is much harder to prove and it has been proven for Standard ML. Important ML features include:

- ML is higher-order supporting functions as first-class values. This means functions may be passed as parameters to functions and returned as values from functions.
- The strong type checking means it is pretty infrequent that you need to debug your code. What a great thing!
- Pattern-matching is used in the specification of functions in ML. Pattern-matching is convenient for writing recursive functions.
- The exception handling system implemented by Standard ML has been proven type safe, meaning that the type system encompasses all possible paths of execution in an ML program.

ML is a very good language to use in learning to implement other languages. It includes tools for automatically generating parts of a language implementation including components called a scanner and a parser which are introduced in chapter 2. These tools, along with the polymorphic strong type checking provided by Standard ML, make implementing a compiler or interpreter a much easier task. Much of the work of implementing a program in Standard ML is spent in making sure all the types in the program are correct. This strong type checking often means that once a program is properly typed it will run the first time. This is quite a statement to make, but nonetheless it is often true.

A Brief History of Prolog

Prolog was developed in 1972 by Alain Colmerauer with Philippe Roussel. Colmerauer and Roussel and their research group had been working on natural language processing for the French language and were studying logic and automated theorem proving[9] to answer simple questions in French. Their research led them to invite Robert Kowalski (pictured in figure 1.8), who was working in the area of logic programming and had devised an algorithm called SL-Resolution, to work with them in the summer of 1971[15][40].

Colmerauer and Kowalski, while working together in 1971, discovered a way formal grammars could be written as clauses in predicate logic. Colmerauer soon devised a way that logic predicates could be used to express grammars that would allow automated theorem provers to parse natural language sentences efficiently. We'll see how this is done in chapter 7.

Fig. 1.7: Alain Colmerauer[8]

In the summer of 1972, Kowalski and Colmerauer worked together again and Kowalski was able to describe the procedural interpretation of what are known as Horn Clauses. Much of the debate at the time revolved around whether logic programming should focus on procedural representations or declarative representations. The work of Kowalski showed how logic programs could have a dual meaning, both procedural and declarative.

Colmerauer and Roussel used this idea of logic programs being both declarative and procedural to devise Prolog in the summer and fall of 1972. The first large Prolog program, which implemented a question and answering system in the French language, was written in 1972 as well.

Fig. 1.8: Robert Kowalski[16]

Later, the Prolog language interpreter was rewritten at Edinburgh to compile programs into DEC-10 machine code. This led to an abstract intermediate form that is now known as the Warren Abstract Machine or WAM. WAM is a low-level intermediate representation that is well-suited for representing Prolog programs. The WAM can be (and has been) implemented on a wide variety of hardware. This means that Prolog implementations exist for most computing platforms.

☞ Practice 1.4

Answer the following questions.

1. Who invented C++? C? Standard ML? Prolog? Ruby?
2. What do Standard ML and Prolog's histories have in common?
3. What language or languages was Ruby based on?

1.4 Language Implementation

There are three ways that languages can be implemented. They can either be

- Interpreted
- Compiled
- Somewhere in between

The goal of each implementation method is to translate a source program into a low-level representation. In other words, to be executable a program must be translated from an English-like format to a format more suitable for a computer to understand. In the case of a compiled language the low-level format is the native machine language of the machine they are going to be executed on. In the case of an interpreted language the low-level representation of a program is a data structure in the memory of the interpreter. The internal data structure representing the program is used by the interpreter to control the execution of the interpreter on the data of the program. The next sections present these implementations in more detail.

Compilation

One method of translating a program to machine language is called compilation. The process is shown in figure 1.9. A compiler is a tool that translates a source program to an assembly language program. Then a tool called an assembler translates the assembly language program to the final machine language program. Machine language is the only type of language that a Central Processing Unit (CPU) understands. The CPU is the *brain* of the computer.

Anyone can write a bad compiler. Good compilers are written using systematic methods that have been developed the last fifty years or so. C, C++, Pascal, Fortran, COBOL and many others are typically compiled languages. On the Linux operating system the C compiler is called *gcc* and the C++ compiler is called *g++*. The *g* in both names reflects the fact that both compilers come out of the GNU project and the Free Software Foundation. Linux, gcc, and g++ are freely available to anyone who wants to download them. The best way to get these tools is to buy or download

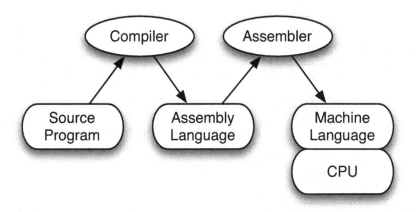

Fig. 1.9: The Compilation Process

a Linux distribution and install it on a computer. The *gcc* and *g++* compilers come standard with Linux.

There are also implementations of gcc and g++ for many other platforms. The web site gcc.gnu.org contains links to source code and to prebuilt binaries for the compiler. You can get versions of the compiler for Windows and many versions of Unix. Since Mac OS is really a Unix platform as well, the gcc compiler is available for Macs too. Use fink to get the compiler for Mac OS. Fink is a tool for downloading many free Unix applications for Mac OS.

Interpretation

Interpreted programs are usually smaller than compiled programs but this is not always true. They can be useful for writing smaller, short programs that you want to try out quickly. Many scripting languages are interpreted. These languages include Ruby, older versions of Basic, Bash, Csh, and others.

When you run a program with an interpreter (see figure 1.10), you are actually running the interpreter. Your program is not running because your program is never translated to machine language. The interpreter is the one big program that all the programs you write in the interpreted language execute. The source program you write controls the behavior of the interpreter program.

One advantage of interpretation is that it encourages programmers to write something and quickly try it out before proceeding. It is often easier to try a piece of a program in an interpreter without writing the entire program. This is called prototyping and it is easier to do with an interpreted language.

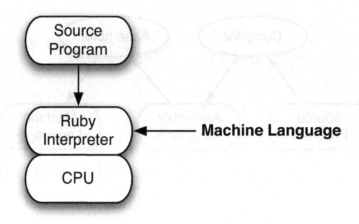

Fig. 1.10: The Interpretation Process

Another advantage is that your source program is portable to different platforms. If you write a program in Ruby you can be pretty sure it will run on Linux, Windows, or Mac OS without any changes. Unless you use a platform specific package in Ruby, a Ruby program will run regardless of the platform. This may be true of compiled programs as well, but they must typically be recompiled for the new platform.

Part of the portability problem revolves around the use of libraries in compiled programs. Libraries are often platform dependent. To move a program from one platform to another the same libraries must exist in a version that is compiled for the new target platform. While POSIX is a standard set of libraries that deals with some of these issues, there are still platform specific libraries that are frequently used. Interpreters generally have a standard set of libraries that come with the system. By standardizing the libraries or implementing them using the interpreted language, developers can be assured that a program implemented using an interpreted language will run the same regardless of the underlying architecture of the machine.

The interpreter itself isn't platform independent. There must be a version of an interpreter for each platform/language combination. So there is a Ruby interpreter for Linux, another for Windows, and yet another for Macs. In fact, with the introduction of Intel based Macs there must be different versions of the Ruby interpreter: one for the PowerPC Macs and another for the Intel based Macs. So, the definition of a platform is an *Operating System/CPU Architecture* combination. Thankfully, the same Ruby interpreter code can be compiled (with some small changes) for each platform.

Yes, you read that last sentence right. The interpreter is a compiled program. There is some piece of every language implementation that is compiled because only programs translated into machine language can run on a CPU.

A huge problem that has driven research into interpreted languages is the problem of heap memory management. Recall that the heap is the place where memory is dynamically allocated. Large C and C++ programs are notorious for having memory leaks. Every time a C++ programmer reserves some space on the heap he/she must remember to free that space. If they don't free the space when they are done with it the space will never be available again while the program continues to execute. The heap is a big space, but if a program runs long enough and continues to allocate and not free space, eventually the heap will fill up and the program will terminate abnormally. In addition, even if the program doesn't terminate abnormally, the performance of the system will degrade as more and more time is spent managing the large heap space.

Many interpreted languages don't require programmers to free space on the heap. Instead, there is a special task or thread that runs periodically as part of the interpreter to check the heap for space that can be freed. This task is called the garbage collector. Programmers can allocate space on the heap but don't have to be worried about freeing that space. For a garbage collector to work correctly space on the heap has to be allocated and accessed in the right way. Many interpreted languages are designed to insure that a garbage collector will work correctly.

The disadvantage of an interpreted language is in speed of execution. Interpreted programs typically run slower than compiled programs. In addition, if an application has real-time dependencies then having the garbage collector running at more or less random intervals may not be desirable. Some language implementations do allow you to control when the garbage collector runs in certain situations, but if the developer is managing when the garbage collector runs in some ways they might as well manage the heap themselves. As you'll read in the next section some steps have been taken to reduce the difference in execution time between compiled and interpreted languages.

Hybrid Language Implementations

The advantages of interpretation over compilation are pretty significant. It turns out that one of the biggest advantages is the portability of programs. It's nice to know when you invest the time in writing a program that it will run the same on Linux, Windows, Mac OS, or some other operating system. This portability issue has driven a lot of research into making interpreted programs run as fast as compiled languages.

Programs written using hybrid languages are compiled. However, the compiled hybrid language program is interpreted. Source programs in the language are not interpreted directly. They are first translated (i.e. compiled) to a lower level language often called a byte-code representation reflecting the fact that it is a low-level representation of the program. This intermediate byte-code form is what is actually interpreted (see figure 1.11). By adding this intermediate step the interpreter can be smaller and faster than traditional interpreters. Sometimes this intermediate step

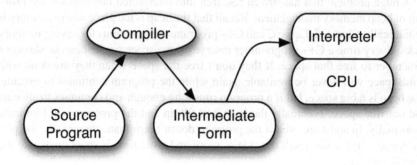

Fig. 1.11: Hybrid Language Implementation

is hidden from the programmer by the programming environment. Sometimes the programmer is required to perform this intermediate step themselves.

Languages that fall into this hybrid category include Java, ML, Python, C#, Visual Basic .NET, JScript, and other .NET platform languages. Both ML and Python include interactive interpreters as well as the ability to compile and run low-level byte-code programs. The byte-code files are named *.pyc* files in the case of Python. In, ML the compiled files are named with a *-platform* as the last part of the compiled file name.

The Java and the .NET programming environments do not include interactive interpreters. The only way to execute programs with these platforms is to compile the program and then run the compiled program using the interpreter. The interpreter is called the Java Virtual Machine in the case of Java. The Java Virtual Machine program is named *java* on all platforms. Programs written for the .NET platform run under Microsoft Windows and in some cases Linux. Microsoft submitted some of the .NET specifications to the ISO to allow third party software companies to develop support for .NET on other platforms. In theory all .NET programs are portable like Java, but so far implementations of the .NET framework are not as generally available as Java. The Java platform has been implemented and released on all major platforms. In fact, in November 2006 Sun, the company that created Java, announced they were releasing the Java Virtual Machine and related software under the GNU Public License to encourage further development of the language and related tools.

The intermediate form under Java is called a Java byte-code file. Byte-code files in Java have a *.class* extension. Under the .NET Framework a byte-code file is called a managed module which includes the Intermediate Language (IL) and metadata. These files are produced by the compilers for each platform and read by their interpreter or virtual machine.

1.5   Where do we go from here?

The next chapter starts by introducing a low-level language called EWE which is an extended version of the RAM language[29]. The chapter will introduce syntax of languages through examples with the EWE language and a small calculator language.

When learning a new language, which you will do many times in your career, it is very helpful to understand something about language implementation. Subsequent chapters in the book will look at language implementation to better understand the languages you are learning, their strengths and weaknesses. While learning these languages you will also be implementing interpreters and/or compilers for some simple languages. This will give you insight into language implementation and knowledge of how to use these languages to solve problems.

While learning new languages and studying programming language implementation it becomes important to understand models of computation. A compiler translates a high-level programming language into a lower level computation. These low-level computations are usually expressed in terms of machine language but not always. More important than the actual low-level language is the model of computation. Some models are based on register machines. Some models are based on stack machines. Still other models may be based on something entirely different. You'll be exploring these models of computation in more detail as you read this text.

1.6 Exercises

1. What are the three ways of thinking about programming, often called programming paradigms?
2. Name at least one language for each of the three methods of programming described in the previous question?
3. Name one person who had a great deal to do with the development of the imperative programming model. Name another who contributed to the functional model. Finally, name a person who was responsible for the development of the logic model of programming?
4. What are the primary characteristics of each of the imperative, functional, and logic models?
5. Who are the main people involved in each of the four languages this text covers: C++, Ruby, Standard ML, and Prolog?
6. Where are the people you mentioned in the previous question today? What do they do now?
7. Why is compiling a program preferred over interpreting a program?
8. Why is interpreting a program preferred over compiling a program?
9. What benefits do hybrid languages have over interpreted languages?

1.7 Solutions to Practice Problems

These are solutions to the practice problems. You should only consult these answers after you have tried each of them for yourself first. Practice problems are meant to help reinforce the material you have just read so make use of them.

Solution to Practice Problem 1.1

1. The run-time stack, global memory, and the heap are the three divisions of data memory.
2. Data on the heap is created at run-time.
3. An activation record holds information like local variables, the program counter, the stack pointer, and other state information necessary for a function invocation.
4. An activation record is created each time a function is called.
5. An activation record is deleted when a function returns.
6. The primary goal of imperative, object-oriented programming is to update memory by updating variables and/or objects as the program executes. The primary operation is memory updates.

Solution to Practice Problem 1.2

1. Functional languages include Standard ML, Lisp, Haskell, and Scheme.
2. In the imperative model the primary operation revolves around updating memory (the assignment statement). In the functional model the primary operation is function application.
3. The functional model emphasizes immutable data. However, some imperative languages have some immutable data as well. For instance, Java strings are immutable.

Solution to Practice Problem 1.3

1. You never write a program in Prolog. You write a database of rules in Prolog that tell the single Prolog program (depth first search) how to proceed.
2. The programmer provides a database of facts and predicates that tell Prolog about a problem. In Prolog the programmer describes the problem instead of programming the solution.

Solution to Practice Problem 1.4

1. C++ was invented by Bjourne Stroustrup. C was created by Dennis Ritchie. Standard ML was primarily designed by Robin Milner. Prolog was designed by Alain Colmerauer and Philippe Roussel with the assistance of Robert Kowalski. Ruby was created by Yukihiro Matsumoto (Matz for short).
2. Standard ML and Prolog were both designed as languages for automated theorem proving first. Then they became general purpose programming languages later.
3. Ruby has many influences, but Smalltalk and Perl stand out as two of the primary influences on the language.

1.8 Additional Reading

The history of languages is fascinating and a lot more detail is available than was covered in this chapter. There are many great resources on the web where you can get more information. Use Google or Wikipedia and search for "History of your_favorite_language" as a place to begin. However, be careful. You can't believe everything you read on the web and that includes Wikipedia. While the web is a great source, you should always research your topic enough to independently verify the information you find there.

Chapter 2

Specifying Syntax

Once you've learned how to program in some language, learning a new programming language isn't all that hard. When learning a new language you need to know two things. First, you need to know what the keywords and constructs of the language look like. In other words, you need to know the mechanics of putting a program together in the programming language. Are the semicolons in the right places? Do you use begin...end or do you use curly braces (i.e. { and }). Learning how a program is put together is called learning the syntax of the language. Syntax refers to the words and symbols of a language and how to write the symbols down in the right order.

Semantics is the word that is used when deriving meaning from what is written. The semantics of a program refers to what the program will do when it is executed. Informally, it may be much easier to say what a program does than to describe the syntactic structure of the program. However, syntax is a lot easier to describe formally than semantics. In either case, if you are learning a new language, you need to learn something about the syntax of the language first.

2.1 Terminology

Once again, **syntax** of a programming language determines the well-formed or grammatically correct programs of the language. **Semantics** describes how or whether such programs will execute.

- **Syntax** is how things look
- **Semantics** is how things work (the meaning)

Many questions we might like to ask about a program either relate to the syntax of the language or to its semantics. It is not always clear which questions pertain to syntax and which pertain to semantics. Some questions may concern semantic issues that can be determined statically, meaning before the program is run. Other semantic issues may be dynamic issues, meaning they can only be determined at run-time. The difference between static semantic issues and syntactic issues is sometimes a difficult distinction to make.

K.D. Lee, *Programming Languages*, DOI: 10.1007/978-0-387-79421-1_2,

Example 2.1

Apparently

```
a=b+c;
```

is correct C++ syntax. But is it really a correct statement?

1. Have b and c been declared as a type that allows the + operation?
2. Is a assignment compatible with the result of the expression b+c?
3. Do b and c have values?
4. Does the assignment statement have the proper form?

There are lots of questions that need to be answered about this assignment statement. Some questions could be answered sooner than others. When a C++ program is compiled it is translated from C++ to machine language as described in the previous chapter. Questions 1 and 2 are issues that can be answered when the C++ program is compiled. However, the answer to the third question above might not be known until the C++ program executes. The answers to questions 1 and 2 can be answered at *compile-time* and are called *static* semantic issues. The answer to question 3 is a *dynamic* issue and is probably not determinable until run-time. In some circumstances, the answer to question 3 might also be a static semantic issue. Question 4 is definitely a syntactic issue.

Unlike the dynamic semantic issues discussed above, the correct syntax of a program is definitely statically determinable. Said another way, determining a syntactically valid program can be accomplished without running the program. The syntax of a programming language is specified by something called a grammar. But before discussing grammars, the parts of a grammar must be defined. A **terminal** or **token** is a symbol in the language.

- C++ terminals: **while, const, (, ;, 5, b**
- Terminal types are keywords, operators, numbers, identifiers, etc.

A **syntactic category** or **nonterminal** is a set of objects (strings) that will be defined in terms of symbols in the language (terminal and nonterminal symbols).

- C++ nonterminals: <statement>, <expression>, <if-statement>, etc.
- Syntactic categories define parts of a program like statements, expressions, declarations, and so on.

A **metalanguage** is a higher-level language used to specify, discuss, describe, or analyze another language. English is used as a metalanguage for describing programming languages, but because of the ambiguities in English, more formal metalanguages have been proposed. The next section describes a formal metalanguage for describing programming language syntax.

2.2 Backus Naur Form (BNF)

Backus Naur Format (i.e. BNF) is a formal metalanguage for describing language syntax. The word *formal* is used to indicate that BNF is unambiguous. Unlike English, the BNF language is not open to our own interpretations. There is only one way to read a BNF description.

BNF was used by John Backus to describe the syntax of Algol in 1963. In 1960, John Backus and Peter Naur, a computer magazine writer, had just attended a conference on Algol. As they returned from the trip it became apparent that they had very different views of what Algol would look like. As a result of this discussion, John Backus worked on a method for describing the grammars of languages. Peter Naur slightly modified it. The notation is called BNF, or Backus Naur Form or sometimes Backus Normal Form. BNF consists of a set of rules that have this form:

<syntactic category> ::= a string of terminals and nonterminals
"::=" means "is composed of " (sometimes written as →)

Often, multiple rules defining the same syntactic category are abbreviated using the "|" character which can be read as "or" and means set union. That is the entire language. It's not a very big metalanguage, but it is powerful. Consider the following examples.

Example 2.2

BNF Examples from Java

<primitive type> ::= boolean
<primitive type> ::= char

Abbreviated

<primitive type> ::= boolean | char | byte | short | int | long | float | ...
<argument list> ::= <expression> | <argument list> , <expression>
<selection statement> ::=
    if (<expression>) <statement>
| if (<expression>) <statement> else <statement>
| switch (<expression>) <block>
<method declaration> ::=
    <modifiers> <type specifier> <method declarator>
    throws <method body>
| <modifiers> <type specifier> <method declarator> <method body>
| <type specifier> <method declarator> throws <method body>
| <type specifier> <method declarator> <method body>

The above description can be described in English as *the set of method declarations is the union of the sets of method declarations that explicitly throw an exception with those that don't explicitly throw an exception with or without modifiers attached to their definitions.* The BNF is much easier to understand and is not ambiguous like this English description.

2.3 The EWE Language

EWE is an extension of a primitive language called RAM designed by Sethi[29] as a teaching language. RAM stands for Random Access Machine. You might ask, "Did you intentionally name this language EWE?". "Yes!", I'd sheepishly respond. You can think of the EWE language as representing the language of a simple computer. EWE is an interpreter much the way the Java Virtual Machine is an interpreter of Java byte codes. EWE is much simpler than the language of the Java Virtual Machine.

Example 2.3

Consider the C++ program fragment.

```
1   int  a=0;
2   int  b=5;
3   int  c=b+1;
4   a=b*c;
5   cout << a;
```

The EWE code below implements the C++ program fragment above.

```
1   a := 0
2   b := 5
3   one := 1
4   c := b + one
5   a := b * c
6   writeInt(a)
7   halt
8   equ a M[0]    b M[1]    c M[2]    one M[3]
```

As you can see, there is a very close correspondence between the C++ program and the EWE program. You can't write **c=b+1** in EWE directly. That required a little extra work. Of course, that's not the only program that might implement the C++ program fragment given above.

Example 2.4

Here's another EWE program that computes the same thing as the C++ program fragment given above. This EWE program isn't quite as straightforward as the last one, but they do the same thing.

```
1   # int a=0;
2   R0:=0      # load 0 into R0
3   M[SP+12]:=R0
4   # int b = 5;
5   R1:=5      # load 5 into R1
6   M[SP+13]:=R1
7   # int c = b+1;
8   R2:=SP         # b+1
9   R2:=M[R2+13]   # load b into R2
```

```
10   R3:=1      # load 1 into R3
11   R2:=R2+R3
12   M[SP+14]:=R2
13   # a = b*c;
14   R4:=SP         # b*c
15   R4:=M[R4+13]    # load b into R4
16   R5:=SP
17   R5:=M[R5+14]    # load c into R5
18   R4:=R4*R5
19   R6:=SP
20   M[R6+12]:=R4
21   R7:=M[SP+12]
22   writeInt(R7)
23   halt
24   equ SP M[10]    equ R0 M[0]    equ R1 M[0]
25   equ R2 M[0]     equ R3 M[1]    equ R4 M[0]
26   equ R5 M[1]     equ R6 M[1]    equ R7 M[0]
```

The EWE language's interpreter recognizes one statement per line. Comments begin with a # and extend to the end of the line. The statements are followed by equates that equate identifiers to memory locations. The EWE computation model consists of:

- data memory locations specified by M[...]
- an instruction memory containing statements

Statements in a EWE program are executed in sequence unless a goto statement is executed. Statement execution terminates when an error occurs or the **halt** statement is executed.

EWE BNF

The syntax of the EWE language is completely specified by the BNF given on page 26. The semantics of the interpreter is not. The `null` symbol is there to draw attention to the fact that the equates part may be empty (there might not be any equates in a program). Keywords are not case sensitive. Strings are delimited by single or double quotes.

The readStr function reads a string and places the first character in the first memref location. It continues putting characters of the string in successive memory locations until either the string ends or the string surpasses the length stored in the second memref minus 1. Strings are terminated with a null (i.e. 0) character. Note that while a single memory location is big enough to hold four characters, only one character is placed in each memory location.

The writeStr function writes a string starting at the memref location and extending in successive memory locations until a null character is encountered. If a null character does not terminate the string, the interpreter will raise an illegal memory reference exception.

```
 1   <eweprog> ::= <executable> <equates> EOF
 2
 3   <executable> ::=
 4         <labeled instruction>
 5       | <labeled instruction> <executable>
 6
 7   <labeled instruction> ::=
 8         Identifier ":" <labeled instruction>
 9       | <instr>
10
11   <instr> ::=
12         <memref> ":=" Integer
13       | <memref> ":=" String
14       | <memref> ":=" "PC" "+" Integer
15       | "PC" ":=" <memref>
16       | <memref> ":=" <memref>
17       | <memref> ":=" <memref> "+" <memref>
18       | <memref> ":=" <memref> "-" <memref>
19       | <memref> ":=" <memref> "*" <memref>
20       | <memref> ":=" <memref> "/" <memref>
21       | <memref> ":=" <memref> "%" <memref>
22       | <memref> ":=" "M" "[" <memref> "+" Integer "]"
23       | "M" "[" <memref> "+" Integer "]" ":=" <memref>
24       | "readInt" "("<memref> ")"
25       | "writeInt" "(" <memref> ")"
26       | "readStr" "("<memref> "," <memref> ")"
27       | "writeStr" "(" <memref> ")"
28       | "goto" Integer
29       | "goto" Identifier
30       | "if" <memref> <condition> <memref> "then" "goto" Integer
31       | "if" <memref> <condition> <memref> "then" "goto" Identifier
32       | "halt"
33       | "break"
34
35   <equates> ::=
36         null
37       | "equ" Identifier "M" "[" Integer "]" <equates>
38
39   <memref> ::=
40         "M" "[" Integer "]"
41       | Identifier
42
43   <condition> ::= ">=" | ">" | "<=" | "<" | "=" | "<>"
```

Listing 2.1: The EWE BNF

☞ Practice 2.1

The following program is not a valid EWE program. Using the BNF for EWE
list the problems with this program.

```
1      readln(A);
2      readln(B);
3      if A-B < 0 then
4          writeln(A)
5      else
6          writeln(B);
```

How could you rewrite this program so that it does what this program intends to
do?

☞ Practice 2.2

Write a EWE program to read a number from the keyboard and print out the
sum of all the numbers from 1 to that number.

Example 2.5

EWE is essentially an assembly language. It contains a few higher-level constructs, but very few. The EWE program given below upper cases all the characters in a string read from the keyboard. The simple way to write an assembly
language program is to first write it in a high-level language. For instance, the
program might look something like this in a C-like language.

```
1   s = input();
2   i = 0;
3   while s[i] != 0 {
4     if ('a' <= s[i] && s[i] <= 'z')
5        s[i] = s[i] - 'a' + 'A';
6     i++;
7   }
8
9   printf("%s",s)
```

When writing the program in EWE you will want to program the opposite of
any if-then or while loop conditions you wrote in the high-level language. This is
because you are going to use a **goto** statement to assist in completing the code. If
the condition is false in an if-then statement you will jump around the **then** part
of the statement by jumping to code that is after the **then** part. The code below
shows you the EWE code with the appropriate C code intermingled as comments.
Comments in EWE begin with a pound sign (i.e. #).

```
1   zero:=0
2   one:=1
3   littlea := 97
4   littlez := 122
5   diff:=32
```

```
 6  # s = input();
 7  len:=100
 8  readStr(s,len)
 9  # i=0;
10  i:=100
11  # while s[i]!=0 {
12  loop: tmp:=M[i+0]
13  if tmp = zero then goto end
14  #    if ('a' <= s[i] && s[i] <= 'z')
15  if littlea > tmp then goto skip
16  if tmp > littlez then goto skip
17  #       s[i] = s[i] - 32;
18  tmp:=tmp-diff
19  M[i+0]:=tmp
20  skip:
21  #    i++;
22  i:=i+one
23  goto loop
24  # printf("%s",s)
25  end: writeStr(s)
26  halt
27
28  equ zero M[0] equ one M[1] equ littlea M[2]
29  equ littlez M[3] equ diff M[4] equ len M[5]
30  equ s M[100] equ tmp M[6] equ i M[7]
```

☞ Practice 2.3

> Write a EWE program that reads a list of numbers from the screen and prints
> them out in reverse order. In order to do this exercise you need to know some-
> thing about indexed addressing (see the example above).
> **HINT:** What kind of data structure lets you reverse the elements of a list?

2.4 Context-Free Grammars

Another name for a BNF grammar is a context-free grammar. The only difference is
in the metalanguage used to write the grammar. A context-free grammar is defined
as a four tuple:

$$G = (\mathcal{N}, \mathcal{T}, \mathcal{P}, \mathcal{S})$$

where

- $\mathcal{N}$ is a set of symbols called nonterminals or syntactic categories.
- $\mathcal{T}$ is a set of symbols called terminals or tokens.
- $\mathcal{P}$ is a set of productions of the form $n \to \alpha$ where n is a nonterminal and α is a string of terminals and nonterminals.
- $\mathcal{S}$ is a special nonterminal called the start symbol of the grammar.

Example 2.6

A grammar for expressions in programs can be specified as $G = (\mathcal{N}, \mathcal{T}, \mathcal{P}, E)$ where

$\mathcal{N} = \{E, T, F\}$
$\mathcal{T} = \{identifier, number, +, -, *, /, (,)\}$
$\mathcal{P}$ is defined by the set of productions
$E \rightarrow E + T \mid E - T \mid T$
$T \rightarrow T * F \mid T/F \mid F$
$F \rightarrow (E) \mid identifier \mid number$

2.5 Derivations

A *sentence* of a grammar is a string of tokens from the grammar. A sentence belongs to the language of a grammar if it can be derived from the grammar. This process is called constructing a derivation. A *derivation* is a sequence of sentential forms that starts with the start symbol of the grammar and ends with the sentence you are trying to derive. A *sentential form* is a string of terminals and nonterminals from the grammar. In each step in the derivation, one nonterminal of a sentential form, call it A, is replaced by a string of terminals and nonterminals, β, where $A \rightarrow \beta$ is a production in the grammar.

While the previous paragraph is a bit dense to read the first time it really isn't that hard. An example should clear things up.

Example 2.7

Prove that the expression (5*x)+y is a member of the language defined by the grammar given in example 2.6 by constructing a derivation for it.

The derivation begins with the start symbol of the grammar and ends with the sentence.

$E \Rightarrow \underline{E + T} \Rightarrow T + T \Rightarrow F + T \Rightarrow \underline{(E) + T} \Rightarrow (T) + T \Rightarrow \underline{(T * F) + T} \Rightarrow (F * F) +$
$T \Rightarrow \underline{(5 * F) + T} \Rightarrow (5 * x) + T \Rightarrow \underline{(5 * x) + F} \Rightarrow \underline{(5 * x) + y}$

The underlined parts are all examples of sentential forms.

☞ Practice 2.4

Construct a derivation for the expression $4 + (a - b) * x$.

Types of Derivations

A sentence of a grammar is **valid** if there exists at least one derivation for it using the grammar. There are typically many different derivations for a particular sentence of a grammar. However, there are two derivations that are of some interest to us in understanding programming languages.

- Left-most derivation - Always replace the left-most nonterminal when going from one sentential form to the next in a derivation.
- Right-most derivation - Always replace the right-most nonterminal when going from one sentential form to the next in a derivation.

Example 2.8

The derivation of the sentence $(5 * x) + y$ in example 2.7 is a left-most derivation. A right-most derivation for the same sentence is:

$$E \Rightarrow E + T \Rightarrow E + F \Rightarrow E + y \Rightarrow T + y \Rightarrow F + y \Rightarrow (E) + y \Rightarrow (T) + y \Rightarrow (T * F) + y \Rightarrow (T * x) + y \Rightarrow (F * x) + y \Rightarrow (5 * x) + y$$

☞ Practice 2.5

Construct a right-most derivation for the expression $x * y + z$.

2.6 Parse Trees

A grammar for a language can be used to build a tree representing a sentence of the grammar. This kind of tree is called a *parse tree* for reasons that will become clear in the next section. A parse tree is another way of representing a sentence of a given language. A parse tree is constructed with the start symbol of the grammar at the root of the tree. The children of each node in the tree must appear on the right hand side of a production with the parent on the left hand side of the same production. A program is syntactically valid if there is a parse tree for it using the given grammar.

While there are typically many different derivations of a sentence in a language, there is only one parse tree. This is true as long as the grammar is not ambiguous. In fact that's the definition of ambiguity in a grammar. A grammar is *ambiguous* if and only if there is a sentence in the language of the grammar that has more than one parse tree. See section 2.11 for more information.

Example 2.9

The parse tree for the sentence derived in example 2.7 is depicted in figure 2.1. Notice the similarities between the derivation and the parse tree.

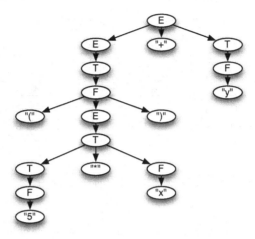

Fig. 2.1: A Parse Tree

☞ Practice 2.6

What does the parse tree look like for the right-most derivation of (5*x)+y?

☞ Practice 2.7

Construct a parse tree for the expression "4+(a-b)*x".
HINT: What has higher precedence, "+" or "*"? The grammar given above auto-matically makes "*" have higher precedence. Try it the other way and see why!

2.7 Parsing

Parsing is the process of detecting whether a given string of tokens is a valid sen-tence of a grammar. Every time you compile a program or run a program in an interpreter the process described in this section is executed. Sometimes it completes successfully and sometimes it doesn't. When it doesn't you are told there is a syntax error in your program. A *parser* is a program that given a sentence, checks to see if the sentence is a member of the language of the given grammar. It may or may not construct a parse tree for the sentence at the same time.

- A top-down parser starts with the root of the tree
- A bottom-up parser starts with the leaves of the tree

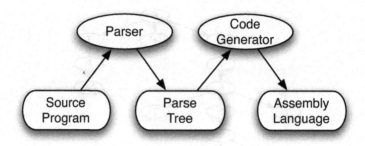

Fig. 2.2: Flow of Data surrounding a Parser

Top-down and bottom-up parsers check to see if a sentence belongs to a grammar by constructing a derivation for the sentence, using the grammar. A parser either reports success (and possibly returns the parse tree) or reports failure (hopefully with a nice error message). The flow of data is pictured in figure 2.2.

2.8 Parser Generators

A parser generator is a program that given a grammar, constructs a parser for the language specified by the grammar. This is a program that generates a program as pictured in figure 2.3. Examples of parser generators are yacc and ml-yacc. They both generate bottom-up parsers.

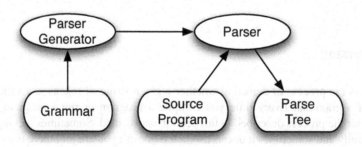

Fig. 2.3: Flow of Data surrounding a Parser Generator

2.9 Bottom-Up Parsers

As described above, bottom-up parsers are generally generated by a parser generator like ml-yacc (used by ML programs) or yacc (used by C and C++ programs). Parser generators construct a parse tree from the bottom up. We can be more specific. They actually construct a reverse right-most derivation of the sentence (i.e. Source program).

A parser generator works by (possibly) looking at the next token (i.e. terminal) in the input and then decides based on that and the partial derivation so far which production to apply to get the next step in the reverse right-most derivation. This algorithm uses a particular type of abstract machine called a push-down automaton. You need a particular kind of grammar to construct a push-down automaton called an LALR(1) grammar. Many grammars are LALR(1). You can learn more about push-down automata in a compiler construction text. It is beyond the scope of this book.

2.10 Top-Down Parsers

Top-down parsers are generally written by hand. They are sometimes called recursive descent parsers because they can be written as a set of mutually recursive functions. A top-down parser constructs a left-most derivation of the sentence (i.e. source program).

A top-down parser operates by (possibly) looking at the next token in the source file and deciding what to do based on the token and where it is in the derivation. To operate correctly, a top-down parser must be designed using a special kind of grammar called an LL(1) grammar.

2.11 Other Forms of Grammars

As a computer programmer you will likely learn at least one new language and probably a few during your career. New application areas frequently cause new languages to be developed to make programming applications in that area more convenient. Java, JavaScript, and ASP.NET are three new languages that were created because of the world wide web. A recent trend in programming languages is to develop domain specific languages. So if you are designing elevator controllers you may be programming in a language that was specially designed for that purpose.

Programming language references almost always contain some kind of reference that describes the constructs of the language. Many of these programming references give the grammar of the language using a variation of a context free grammar. A few examples of these grammar variations are given here to make you aware of notation that is often used in language references.

CBL (Cobol-like) Grammars

These were originally used in the description of Cobol. They are not as formal as BNF.

1. Optional elements are enclosed in brackets: [ ].
2. Alternate elements are vertically enclosed in braces: { }.
3. Optional alternates are vertically enclosed in brackets.
4. A repeated element is written once followed by an ellipsis: ...
5. Required key words are underlined; optional noise words are not.
6. Items supplied by the user are written as lower case or as syntactic categories from which an item may be taken.

Example 2.10

Here is the description of the COBOL ADD statement.

$<$Cobol Add statement$>$ ::=

$$\text{ADD} \begin{Bmatrix} \text{identifier} \\ \text{number} \end{Bmatrix} \begin{bmatrix} \text{, identifier} \\ \text{, number} \end{bmatrix} \dots \underline{\text{TO}}$$

identifier [ROUNDED][, identifier [ROUNDED]] ...
[ON SIZE ERROR $<$statement$>$]

One such add statement might be:

ADD A, 5 TO B ROUNDED, D
ON SIZE ERROR PERFORM E-ROUTINE

Extended BNF (EBNF)

Since a BNF description of the syntax of a programming language relies heavily on recursion to provide lists of items, many definitions use these extensions:

1. **item?** or **[item]** means item is optional.
2. **item*** or **{item}** means to take zero or more occurrences of an item.
3. **item+** means to take one or more occurrences of an item
4. Parentheses are used for grouping

Example 2.11

Here is can example of method declarations in Java.

$<$method declaration$>$ ::=
$<$modifiers$>$? $<$type specifier$>$
$<$method declarator$>$ throws ? $<$method body$>$

Syntax Diagrams

A syntax diagram is a graph or graphs that have been used to describe Pascal and other programming languages.

1. A terminal is shown in a circle or oval.
2. A syntactic category is placed in a rectangle.
3. The concatenation of two objects is indicated by a flowline.
4. The aternation of two objects is shown by branching.
5. Repetition of objects is represented by a loop.

Example 2.12

Here are some descriptions of simple expressions in Pascal. Each of these different methods describe the same simple expressions in Pascal. Notice that some descriptions are more compact than the BNF. Each of them are unambiguous in their descriptions.

While BNF is less compact, it is the easiest to enter on a keyboard and for computer programs to read. There is a trade-off between computer readability and human readability that is at the center of many of our decisions about how to formally define programming languages.

BNF

```
<simple expr> ::=
              <term>
        |     <sign> <term>
        |     <simple expr> <adding operator> <term>
<sign> ::= "+" | "-"
<adding operator> ::= "+" | "-" | "or"
```

CBL

$<simple expr> ::=$

$$\begin{bmatrix} + \\ - \end{bmatrix} <term> \begin{bmatrix} \begin{Bmatrix} + \\ - \\ or \end{Bmatrix} <term> \end{bmatrix} \ldots$$

EBNF

$<simple expr> ::= [<sign>] <term> \{<adding operator> <term>\}$

Syntax Diagram

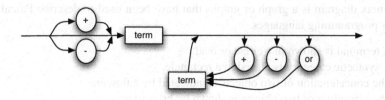

☞ Practice 2.8

According to the syntactic specification in example 2.12, which of these terminal strings are simple expressions, assuming that a, b, and c are legal terms:

1. a+b-c
2. -a or b+c
3. b - - c

Ambiguous Grammars

As stated above, a grammar is ambiguous if there exists more than one parse tree for a given sentence of the language.

Example 2.13

The classic example is nested if-then-else statements. Consider the following Pascal statement:

```
1      if a<b then
2          if b<c then
3              writeln("a<c")
4      else
5          writeln("?")
```

Which *if* statement does the *else* go with? It's not entirely clear. According to the grammar, it could go with either. This means there is some ambiguity in the grammar for Pascal. This resolved by deciding the *else* should go with the nearest *if*. In a bottom-up parser this is called a shift/reduce conflict. In this case it is resolved by shifting instead of reducing.

☞ Practice 2.9

Consider the expression grammar

<expr> ::= identifier | <expr> <operator> <expr>
<operator>::= "+" | "*"

Consider the terminal string a * b + c.
Give two parse trees for this expression. This ambiguity could be resolved by specifying a precedence of operators in the grammar. However, there are better methods than specifying precedence. Precedence of operators can also be specified by introducing extra productions. See example 2.6 on page 29 for a better way of writing the grammar for this language.

2.12 Abstract Syntax Trees

There is a lot of information in a parse tree that isn't really needed to encapsulate the program that it represents. An abstract syntax tree is like a parse tree except that non-essential information is removed. More specifically,

- Nonterminal nodes in the tree are replaced by nodes that reflect the part of the sentence they represent.
- Unit productions in the tree are collapsed.

Example 2.14

For example, the parse tree from figure 2.1 on page 31 can be represented by the following abstract syntax tree.

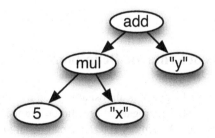

This tree eliminates all the unnecessary information and leaves just what is essential for evaluating the expression. Abstract syntax trees are used by compilers while generating code and by interpreters when running your program. Parse trees are usually not built by the parser, but the parser still constructs a derivation to check the syntax of a program. Usually, at the same time the abstract syntax tree is built.

☞ Practice 2.10

What does the abstract syntax tree of 4+(a-b)*x look like?

2.13 Infix, Postfix, and Prefix Expressions

The abstract syntax tree in example 2.14 represents a computation. We can recover the infix expression it represents by doing an inorder traversal of the abstract syntax tree. To recall, an inorder traversal operates as follows:

```
1  Inorder_traverse(t a tree)
2    If t is an empty tree, do nothing
3    inorder_traverse(left subtree of t)
4    print the data of the root node in the tree t
5    inorder_traverse(right subtree of t)
```

☞ Practice 2.11

Assume there is a BTNode class in your favorite object-oriented language with appropriate constructors, and getData, getLeft, and getRight member functions which return the data at a node, the left subtree, and the right subtree respectively. Write some code to implement this inorder traversal of a tree. Assume the AST in example 2.14 is given as input. What is the output? Is there anything wrong?

☞ Practice 2.12

How does this code change to do a postorder traversal? What is the output given the tree in example 2.14.

2.14 Limitations of Syntactic Definitions

The concrete syntax for a language is almost always an incomplete description. Not all terminal strings generated are regarded as valid programs. For instance, consider the EWE BNF on page 26. A memory reference can be an identifier. The identifier must be defined in an equ statement. But, there is nothing in the grammar specifying this relationship.

In fact, there is no BNF (or EBNF or Syntax Diagram) grammar that generates only legal EWE programs. The same is true for C++, Java, ML, and all programming languages. A BNF grammar defines a context-free language: the left-hand side of each rules contains only one syntactic category. It is replaced by one of its alternative definitions regardless of the context in which it occurrs. The set of programs in any interesting language is not context-free.

Context-sensitive features may be formally described as a set of restrictions or context conditions. Context-sensitive issues deal mainly with declarations of identifiers and type compatibility.

Example 2.15

These are all context-sensitive issues.

- In an array declaration in C++, the array size must be a nonnegative value.
- Operands for the && operation must be boolean in Java.
- In a method definition, the return value must be compatible with the return type in the method declaration.
- When a method is called, the actual parameters must match the formal parameter types.

2.15 Exercises

1. What does the word syntax refer to? How does it differ from semantics?
2. What is a token?
3. What is a nonterminal?
4. What does BNF stand for? What is its purpose?
5. Describe what the rules in lines 35-37 of the EWE BNF on page 26 mean. Answer this in some detail. Saying they define equates is not enough.
6. According to the EWE BNF, how many labels can an instruction have?
7. Given the grammar in example 2.6, derive the sentence (4+5)*3.
8. Draw a parse tree for the sentence (4+5)*3.
9. What kind of derivation does a top-down parser construct?
10. What would the abstract syntax tree for (4+5)*3 look like?
11. Describe how you might evaluate the abstract syntax tree of an expression to get a result? Write out your algorithm in English that describes how this might be done.
12. List four context-sensitive conditions in your favorite language.
13. Write a EWE program that prompts the user to enter three numbers and prints the max of the three numbers to the screen. Think about this before attempting to write it. It might be harder than you think at first.
14. Write a EWE program that prompts the user to enter a string and prints the reverse of that string to the screen.
15. Write a EWE program that prompts the user to enter a string and prints the string back to the screen with the first letter of each word upper cased.
16. Write a EWE program that asks the user to enter a number and prints either the square root of the number if it is an integer or the two integers the square root falls between if it is not an integer result. EWE does not operate on real numbers. It only works with integers and strings.
17. Using the EWE interpreter, write a program that prompts the user for a number and prints the factorial of that number.

2.16 Solutions to Practice Problems

These are solutions to the practice problems. You should only consult these answers after you have tried each of them for yourself first. Practice problems are meant to help reinforce the material you have just read so make use of them.

Solution to Practice Problem 2.1

Here is a correct version of the program. As you can see there are several things wrong with the original.

```
1    readInt(A)
2    readInt(B)
3    C := A - B
4    zero := 0
5    if C >= zero then goto pastwrtA
6        writeInt(A)
7        goto end
8    pastwrtA:
9        writeInt(B)
10   end:
11   halt
12   equ A M[0]    equ B M[1]    equ C M[2]    equ zero M[3]
```

Solution to Practice Problem 2.2

The easiest way to write EWE programs is to write in a language like Java or Python and then translate the code to EWE. Reverse any relational operators to make the translation (see the previous exercise). So for instance, a less than operator becomes greater or equal when translated into EWE. Here is a Python version of the program.

```
1  n = input("Enter a postive integer:")
2  sum = 0
3  for x in range(n+1):
4     sum = sum + x
5
6  print "The sum is", sum
```

And here is a EWE version.

```
1  readInt(n)
2  sum := 0
3  one := 1
4  x := 1
5  loop:
6     if x > n then goto end
7        sum := sum + x
8        x := x + one
```

```
9      goto loop
10   end:
11   writeInt(sum)
12   halt
13   equ sum M[0]    equ one M[1]    equ x M[2]    equ n M[3]
```

If you think hard about this problem there is a simpler version that is about three lines long. You have to find the formula that computes the sum of the first *n* integers, though.

Solution to Practice Problem 2.3

You need to use indexed addressing to create a stack.

```
1    SP := 100
2    hundred := 100
3    zero := 0
4    one := 1
5    readloop:
6      readInt(x)
7      if x = zero then goto printloop
8      M[SP+0] := x
9      SP := SP + one
10     goto readloop
11   printloop:
12     SP := SP - one
13     if SP < hundred then goto end
14     x := M[SP+0]
15     writeInt(x)
16     goto printloop
17   end:
18     halt
19   equ SP M[0]    equ hundred M[1]    equ x M[3]
20   equ zero M[4]    equ one M[5]
```

Solution to Practice Problem 2.4

This is a left-most derivation of the expression.
$$E \Rightarrow E+T \Rightarrow T+T \Rightarrow F+T \Rightarrow 4+T \Rightarrow 4+T*F \Rightarrow 4+F*F \Rightarrow 4+(E)*F \Rightarrow$$
$$4+(E-T)*F \Rightarrow 4+(T-T)*F \Rightarrow 4+(F-T)*F \Rightarrow 4+(a-T)*F \Rightarrow$$
$$4+(a-F)*F \Rightarrow 4+(a-b)*F \Rightarrow 4+(a-b)*x$$

Solution to Practice Problem 2.5

This is a right-most derivation of the expression.
$E \Rightarrow E+T \Rightarrow E+F \Rightarrow E+z \Rightarrow T+z \Rightarrow T*F+z \Rightarrow T*y+z \Rightarrow F*y+z \Rightarrow x*y+z$

Solution to Practice Problem 2.6

Exactly like the parse tree for any other derivation of (5*x)+y. There is only one parse tree for the expression given this grammar.

Solution to Practice Problem 2.7

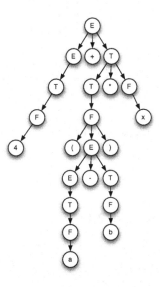

Fig. 2.4: The parse tree for practice problem 2.7

Solution to Practice Problem 2.8

1. a+b-c is a valid simple expression.

2. -a or b + c is a valid simple expression.

3. b - - c is not a simple expression.

Solution to Practice Problem 2.9

In this problem we have a choice of putting the * or the + operator closer to the top of the tree. This will give us two different trees depending on which we choose.

Solution to Practice Problem 2.10

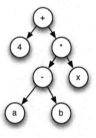

Fig. 2.5: The parse tree for practice problem 2.10

Solution to Practice Problem 2.11

```
1  void inordertraverse(BTNode root) {
2    if (root == nil) then return;
3
4    inordertraverse(root.getLeft());
5    System.out.println(root.getData()+ " ");
6    inordertraverse(root.getRight());
7
8  }
```

The output would be 5 + x * y. The traversal has thrown away the parentheses. If parens are needed the inorder traversal code could be modified to produce a fully parenthesized expression.

Solution to Practice Problem 2.12

The println statement would move to the last line of the function. The postorder output would be 5 x + y *. No parens are needed in a postfix expression.

2.17 Additional Reading

This chapter introduces you to programming language syntax and reading syntactic descriptions. This is a worthwhile skill since you will undoubtedly come across new languages in your career as a computer scientist. There is certainly more that can be said about the topic of syntax of languages. Aho, Sethi, and Ullman [1] have written the widely recognized book on compiler implementation which includes material on syntax definition and parser implementation. There are many other good compiler references as well. The Chomsky hierarchy of languages is also closely tied to this topic. Many books on Discrete Structures in Computer Science introduce this topic and a few good books explore the Chomsky hierarchy more deeply including an excellent text by Peter Linz [21].

Chapter 3

Object-Oriented Programming with C++

This chapter introduces object-oriented programming using C++ through the development of an interpreter for a calculator language. Object-oriented languages are imperative languages employing the imperative model described in section 1.2. However, there is some additional structure introduced in the object-oriented model. Programmers are given tools to organize data and code into objects.

By organizing code and data into objects using *classes*, code reuse is emphasized. Ideas such as *inheritance* enable *polymorphism* among related objects. These terms will be illustrated with examples during the implementation of the intepreter.

The ideas behind object-oriented programming date back to the nineteen seventies and earlier. In the late seventies Modula-2 was one of the first object based languages. However, Modula-2 lacked some of the features of more current object-oriented languages, such as inheritance.

As mentioned earlier, Bjarne Stroustrup was developing C++ during the early eighties. He designed the language to be backward compatible with C so there were some decisions already made for him like the need for *separate compilation* and the presence of a *macro processor*. C++ is one of the most widely used object-oriented languages today.

The use of C/C++ doesn't come without some problems though. The main problem with C/C++ programs are memory leaks. C/C++ programmers must be disciplined in their allocation and deallocation of memory. For some projects, this may be difficult. It is common that programs that run for a long time will have a memory leak that has to be tracked down, which is a difficult task. The problem is that a garbage collector cannot safely be included as part of the model of computation available to C/C++ programs. Both C and C++ were designed to give the programmer maximum control over how the code they wrote was compiled. This meant that more responsibility was left to the programmer and as a result programmers need to be very disciplined when using C/C++.

Modern languages like Java and Ruby provide garbage collection as part of the underlying model of computation. They can do this because these languages are careful about how pointers are exposed to the programmer. In fact pointers are called

K.D. Lee, *Programming Languages*, DOI: 10.1007/978-0-387-79421-1_3,
© Springer Science+Business Media, LLC 2008

references in these languages to distinguish them from pointers in C and C++. The trade-off is that these languages take some control away from the programmer.

Garbage collection does have its own problems. Languages like Java and Ruby aren't as suited to real-time applications where timing is critical. In these languages you never know when garbage collection will occur. Of course there is research into these problems to solve this issue for real-time systems but in general C++ is probably a better choice for real-time applications.

The interpreter developed in this chapter is not a real-time application. But, C++ is chosen because there are many interesting aspects to C++ that can be explored by using it. This chapter is not a complete tutorial on using C++, but will cover enough of the language to develop this project while introducing you to many important concepts in programming languages. Some of the interpreter's code is presented as examples throughout the chapter. While stopping short of a full tutorial on C++, there is enough C++ presented in the examples in this chapter to show you how C++ programs are written, structured, and compiled. Parts of the interpreter are left as a programming project. By the end of this chapter you should have enough background to finish the implementation.

3.1 Application Development

Before beginning to program using C++ it will be helpful to discuss the application development process in light of using C++ as the implementation language. The goal of this chapter is to introduce the design and implementation of a calculator interpreter. C++ code must be compiled before you can run the resulting program. Each of the sections below will go into more detail about how to do each of these tasks using C++ as the implementation language.

Design the Calculator

Let us call the calculator language interpreter *calc*. After completing the interpreter it will be invoked by typing *calc* from a command-line prompt. It should prompt the user to enter a calculator expression at which point the user can enter an expression that will be interpreted. Variations on this interpreter can be implemented including the ability to interpret more than one calculator expression in a session. The calculator will have a single memory location in which an integer value can be stored and recalled. As presented in this chapter the calculator is an integer calculator only. It will do integer division only which means the fractional part is discarded as a remainder after doing division.

Example 3.1

Here is a typical session with the completed calc interpreter.

```
%>calc
Please enter a calculator expression: (4S+5)*(7-R)
The result is 27
%>
```

The S represents the store operator. It stores the value that is computed to the left of it. In this case it stores a 4 in the memory location. The R represents the value that is stored in the memory location.

☞ Practice 3.1

Evaluate the following calculator expressions.

1. (4+5)S*R
2. 3S + R
3. R + 3S
4. 2S*4 + R

While it is certainly possible to build the interpreter in an ad-hoc fashion, the goal of this project is to design the interpreter using a structured approach that will illustrate some of the concepts used in language implementation. In addition, some of the features of C++ will be highlighted as the application is developed.

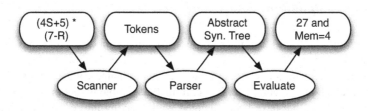

Fig. 3.1: Data flow through the Calc Interpreter

The calculator will read an expression from the command-line and process it as pictured in figure 3.1. The string read from the command-line is fed to a scanner which produces a tokenized representation of the string. The tokens are given to the parser which builds an abstract syntax tree. The abstract syntax tree is then evaluated to produce the result.

The design of the code will closely follow the design in figure 3.1. Because we are using an object-oriented language, the first step is to identify the objects in the figure. There is a scanner, a parser, tokens, and an abstract syntax tree. The only thing left out is evaluate. Evaluate is something we do *to* the abstract syntax tree so we won't define that as an object.

There will also be one more object, a Calculator object that will start the evaluation of an expression. The calculator will have the memory location that we store numbers to and recall numbers from as part of its state.

Each of the objects identified above will be described by a class in C++. The following sections contain examples of the design and implementations for some of these classes. Along the way features of C++, and object-oriented languages in general, are described.

Compiling C++ Programs

C++ is a compiled language. That means we cannot execute a source program directly. It must be compiled to a target language. Typically, there is a C++ compiler for each operating system/architecture combination. That means there is a C++ compiler for Windows/Intel, Mac OS X/PowerPC, Mac OS X/Intel, Linux/Intel, etc. Sometimes there is more than one C++ compiler for a particular operating system/architecture. For instance, Windows has C++ compilers from Microsoft, Borland, the GNU Project and probably other sources as well. The GNU C++ compiler is available for many platforms and is free. It is the compiler that the examples in this book were compiled with.

Separate Compilation

Regardless of which C++ compiler you use, like many modern languages, C++ is organized so that programs may be separately compiled. Without separate compilation, when a project gets extremely large a small change can cause a recompile that could take hours to complete. It is extremely desirable to recompile only those pieces of a project that have been changed while leaving the rest of the compiled code untouched.

Each piece of a C++ project is stored in a separate file. This allows multiple programmers to work on different files at the same time. It also isolates each class in C++ from other class implementations.

The calc interpreter contains six different modules. Five of them, ast.C, calculator.C, parser.C, scanner.C, and token.C contain the implementations of the classes defined for the project. The calc.C module contains the code that ties it all together.

Each module can be compiled separately. To compile a C++ module the command

```
g++ -g -c module.C
```

is executed. The *g++* is the name of the C++ compiler and may differ depending on the compiler you are using. The *-g* tells the compiler to include debug information. The *-c* option tells the compiler to produce an object file. Object files have a *.o* extension. So this command would produce a file called module.o.

When each module has been compiled, then they may be linked together to produce an executable program. That is done with the compiler again using.

```
g++ -o executable_program module1.o module2.o module3.o ...
```

Example 3.2

To completely compile the calc project the following compile commands must be issued.

```
1  g++ -g -c calc.C
2  g++ -g -c scanner.C
3  g++ -g -c token.C
4  g++ -g -c ast.C
5  g++ -g -c parser.C
6  g++ -g -c calculator.C
7  g++ -g -o calc calc.o scanner.o token.o ast.o parser.o \
8     calculator.o
```

A complete compile doesn't have to be done very often. If only one module changes, for instance parser.C, then parser.C would be recompiled and the linking step in the last line would be executed again.

Separate compilation poses some challenges to how a program is organized. For instance, both the parser and the scanner need access to the Token class so they can be compiled. That's the topic of the next section.

Header Files

A header file in C++ is where declarations go that are to be shared between modules. The Token class declaration is needed by the implementation of the Token class located in token.C. It is also needed in the Parser and Scanner classes in parser.C and scanner.C, respectively. The declaration of the Token class does not contain any code and typically header files do not contain code unless it is a very short snippet.

When a program is compiled, the macro processor runs first to expand any macro processor directives in the source code as pictured in figure 3.2. The macro processor looks at all the preprocessor directives and builds the expanded source program that is actually given to the compiler. Every macro processor directive starts with a pound sign (i.e. #).

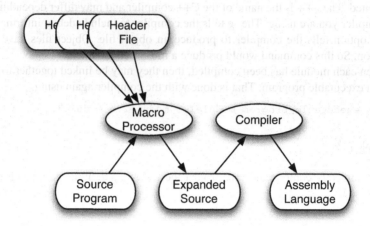

Fig. 3.2: The Macro Processor

Example 3.3

Including a header file in a module is as simple as writing

```
#include "header_file"
```

By convention, header files are named with a .h extension. So there is a header file called scanner.h, token.h, ast.h, parser.h, calculator.h, and a few others in the calculator project containing the declarations of the classes that are a part of the *calc* interpreter.

The parser.C and scanner.C modules include the token.h header file by writing

```
#include "token.h"
```

Sometimes header files depend on declarations that are in other header files. It is legal for one header file to include another header file. The include directive works wherever it is placed.

There are two problems that can result from header files including header files. A declaration can only appear once in the expanded source when compiling. So if you write a module that can include a header file *a.h* and *b.h*, and *b.h* includes *a.h* then you have the declarations in *a.h* declared twice. The compiler will signal an error in that case.

In addition, if *a.h* includes *b.h* which includes *c.h* which includes *a.h* then there is a circular reference created and the macro processor would eventually create an expanded source file that was too large (or some other strange error could occur).

Fortunately, both these problems are solved by using a macro processor directive called *#ifndef*. This directive asks if an identifier is NOT defined. Another macro processor director called *#define* allows you to define an identifier. Identifiers don't need values to be defined. They may be defined or undefined to the macro processor.

Example 3.4

To solve the problem of circular includes or repeated includes the convention is adopted to start include files with an *#ifndef* and then on the next line define the identifier. So the token.h file begins with

```
#ifndef token_h
#define token_h

// Any declarations go here.

#endif
```

The *#ifndef* must be terminated at the end of the include with an *#endif* to end the *#ifndef*. The directive says that if *token_h* is not defined, then on the next line define it and continue including the declarations found in the include file. However, if *token_h* has already been defined it was becuase the include token.h was already included in the expanded source earlier. If this is the case, then the macro processor jumps to the matching *#endif* at the end of the include which causes it to skip all declarations found in the header file. The entire token.h header file is given by taking the code in examples 3.9 and 3.10 and puting the *#ifndef* *#define* macro processor directives before it and the *#endif* directive after it.

Every header file in a C++ project must follow this convention for it to work. C++ programmers are religious about this. You won't see a well written C++ (or C) header file without these directives.

The Make Utility

Executing all the *g++* commands given in example 3.2 would be very tedious if you had to do it more than once. When working on a project you may have to do these commands or some subset of them hundreds of times. Even remembering which subset are required after making a few changes would be a challenge.

Fortunately, there is a utility called *make* that will take care of our housekeeping for us. The make utility takes a file, usually called *Makefile*, as input that specifies the dependencies of a project. A make file is a list of rules. Rules start with the output from the rule followed by a colon. After the colon come the rules or files the output is dependent on. The next line of a rule gives a command to be executed to bring the output of the rule up to date.

The list of rules specify a directed graph of the dependencies between modules of a program. There can be as many rules as you like. Typically there is one rule for each output you expect from the compiler.

Example 3.5

Here is the Makefile for the calc project.

```
1  calc: calc.o scanner.o token.o ast.o parser.o \
2             calculator.o
3      g++ -g -o calc calc.o scanner.o token.o \
4             ast.o parser.o calculator.o
5  calc.o: calc.C scanner.h token.h
6      g++ -g -c calc.C
7  calculator.o: calculator.C calculator.h parser.h ast.h
8      g++ -g -c calculator.C
9  scanner.o: scanner.C scanner.h token.h
10     g++ -g -c scanner.C
11 token.o: token.C token.h
12     g++ -g -c token.C
13 ast.o: ast.C ast.h
14     g++ -g -c ast.C
15 parser.o: parser.C parser.h
16     g++ -g -c parser.C
17 clean:
18     rm -f *.o
19     rm -f calc
```

In lines 1 and 2 (one line in the Makefile) the calc program is dependent on all the object files. The command to bring calc up to date is the g++ linking command. Note that the backslash (i.e. \) is used to extend a line in a makefile.

The words *up to date* have been used pretty liberally in this section. Make actually uses dates to see if everything is *up to date*. Every file has a date attribute, the time the file was last modified. Make looks at dates to determine for instance if the date of calc is older than calc.o. If it is, then the calc rule must be triggered. But, before the linking command is executed, the rules corresponding to calc.o and all the other dependencies listed in the first two lines of the Makefile are examined to see if they should be triggered. Eventually make will arrive at a file that is *up to date* but which is not dependent on anything else.

Example 3.6

To use the make utility you create a file like the one given in example 3.5 and execute the make command

```
make
```

from the directory that contains the Makefile.

☞ Practice 3.2

Assume that you issued the make command to bring everything up to date. Then you change the ast.h header file. Which compile commands will be executed given the Makefile in example 3.5?

Example 3.7

Good Makefiles are hard to write. They almost always have errors in them and the one in example 3.5 is not completely correct. To deal with this there are tools that will generate make files for you. When all else fails you can add an extra clean rule to start over. If something isn't working you can start over by typing

```
make clean
make
```

3.2 The Token Class

The calculator language has several types of tokens. The complete list is *number,+,-,*,/,(,),S,*and *R*. The token class will need to allow each of these tokens to be described. To begin, the type of token can be desribed using something called an enum in C++. An enum is a way that integer values can be given names to make code more readable. Enums are used when you need to enumerate a list of possible values which have some meaning.

Example 3.8

Here is the enum for the types of tokens in the intepreter's language.

```
1   enum TokenType {
2       identifier,keyword,number,add,sub,times,divide,
3       lparen, rparen,eof,unrecognized
4   };
```

The *keyword* token type will be used for both *S* and *R*. The *identifier* and *unrecognized* token will only be used in case of incorrect input. There are no *identifiers* in the calculator language. The *eof* token is returned as the last token. The *identifier*, *keyword*, and *number* tokens each represent a specific identifier, keyword, or number, respectively. This is described in more detail on page 56.

☞ Practice 3.3

Identify the tokens in these expressions. Refer to the *enum* above to be sure you find them all.

1. 3S + R
2. (4+5)S*R

A token has a type. It should also have some extra information that will help should there be an error in the program or expression being evaluated. The line and column where the token began will also be stored with the token. The line will

always be one in this program since the expression is one line long. We have enough
information to specify what a Token object should look like.

Example 3.9

This is the declaration of the Token class.

```
1   class Token {
2   public:
3     Token();
4     Token(TokenType typ, int line, int col);
5     virtual ~Token();
6     TokenType getType() const;
7     int getLine() const;
8     int getCol() const;
9
10  private:
11    TokenType type;
12    int line,col;
13  };
```

The code in example 3.9 is a class declaration in C++. A class declaration typically goes in a header file of the same name. So this class declaration is written in
token.h. The parts of the class declaration are:

1. The class name on line 1.
2. The keyword public identifies a section that contains all public methods and instance variables. Typically instance variables are not public.
3. Line 3 and 4 are the declarations of the Token constructors. Notice that no code
   appears in the class declaration.
4. Line 5 is the Token destructor. This is discussed in more detail below.
5. Lines 7-9 declare accessor methods. The getLex() accessor method is declared
   constant (i.e. *const*) and virtual. We'll discover what those keywords mean soon.
6. The private section is where the instance variables are declared and anything else
   that should be hidden from users of the class.

A C++ class declaration contains no code. It only describes what objects of the
class look like. The Token class described above works well for most tokens. However if a number, keyword, or identifier is discovered in a expression then it is necessary to know what number, keyword, or identifier was found. In this case the
lexeme, or word, of the token is needed. Inheritance is used to extend the Token description to a LexicalToken description. Both classes contain a getLex() method to
be called to get the lexeme from the token. The Token getLex() method only returns
the empty string since lexeme's aren't needed in general, but if the Token is really a
LexicalToken then the getLex() method will return the actual token's lexeme.

Example 3.10

This is the declaration of the LexicalToken class.

```
1   class LexicalToken: public Token {
2   public:
3       LexicalToken(TokenType typ, string* lex,
4                       int line, int col);
5       ~LexicalToken();
6       virtual string getLex() const;
7   private:
8       string* lexeme;
9   };
```

The first line declares the class and indicates that it inherits from the Token class through public inheritance. Public inheritance means the derived class, LexicalToken, has access to the public instance variables and methods of the base class, Token, but not the private members of Token. This is the usual way inheritance is handled. The other line to note is line 6. In this line the getLex() method is declared again except this time it is not declared virtual. Read section 3.4 to learn more about virtual methods.

Any code that wishes to use the Token or LexicalToken classes must have these class declarations within them. The next section will describe how this is accomplished. A lot of unexplained information lies in these two class declarations. The following sections should start to clear up some of the questions you may have. We'll begin by looking at separate compilation.

3.3 Implementing a Class

To implement a class you must write a module with the class implementation in it. Unlike Java, the class declaration and implementation are separate.

Example 3.11

Here is the implementation of the Token and LexicalToken classes contained in the token.C module.

```
1   #include "token.h"
2
3   Token::Token() :
4       type(eof),  line(0),  col(0)
5   {}
6
7   Token::Token(TokenType typ, int lineNum, int colNum) :
8       type(typ),
9       line(lineNum),
10      col(colNum)
11  {}
12
13  Token::~Token() {}
14
15  TokenType Token::getType() const { return type; }
```

```
16
17   int Token::getLine() const { return line; }
18
19   int Token::getCol() const { return col; }
20
21   LexicalToken::LexicalToken(TokenType typ, string* lex,
22                              int lineNum, int colNum) :
23      Token(typ,lineNum,colNum),
24      lexeme(lex)
25   {}
26
27   LexicalToken::~LexicalToken() {
28      try {
29         delete lexeme;
30      } catch (...) {}
31   }
32
33   string LexicalToken::getLex() const {
34      return *lexeme;
35   }
```

Each method of a class is implemented as a function in C++. The *Token::* and *LexicalToken::* that appear before each method name are scope qualifiers. They say that the method being defined appears in the Token or LexicalToken classes. The following sections go into more detail to explain the various parts of the implementation.

Constructors and Initialization Lists

Lines 3-11 contain the constructor implementations for Token. Between the colon and the left brace (i.e. {) appears an initialization list. The list initializes instance variables to values. The value of each instance variable is surrounded by parens. Initialization lists aren't needed in this example. The code could have been written as shown below.

Example 3.12

Another way to write the constructor.

```
1   Token::Token() {
2      type = eof;
3      line = 0;
4      col = 0;
5   }
```

The difference between assignment of instance variables presented in example 3.12 and the initialization list version presented above is minor. Typically, C++

programmers use initialization lists for constructors because the C++ compiler can handle them a little more efficiently when instance variables are themselves objects.

Constructors are never inherited so the LexicalToken constructor calls the Token constructor in its initialization list to initialize the Token part of the object in line 23 of example 3.11. The additional instance variable, lexeme, stores the pointer to the string representation of the token.

Destructors

The declaration

```
string* lexeme;
```

declares a string pointer called lexeme. The star (i.e. *) tells the compiler this a pointer to a string. Pointers in C and C++ are similar to references in Java or Python.

When an object like a LexicalToken stores a pointer to another object, in this case a string, it is likely that the referenced object (the string) is stored on the heap. When the first object is deleted it must be sure to delete the second from the heap. This is the purpose of the code in lines 27-31 of example 3.11. A *destructor* is named the same as the class with a tilde (i.e. ~) in front. The ~ means *not* in some logic notations, so ~LexicalToken means *not* LexicalToken or in other words the destructor. Right before a LexicalToken is deleted the destructor will be called to delete the storage it refers to on the heap. It's a good idea to enclose the deletion code in a try catch block to avoid terminating the program if the data is already deleted for some reason. The ... in the try catch block will catch all exceptions that might occur.

3.4   Inheritance and Polymorphism

When a token is discovered by the scanner it is created and returned as a Token object. The getToken method of the scanner class returns a *Token** (see below). However, some tokens are *Tokens* and some are *LexicalTokens*. Inheritance enables the scanner to simply return either type of token as a *Token**. However, later when the parser is done with the token it will want to delete the token from the heap. Here lies the problem. Is the token being deleted a *Token* or a *LexicalToken*? To know, the parser might look at the type of token its token pointer is pointing at.

But this is bad practice and in this case impossible. The type of an object isn't stored anywhere in a C++ program (although Token objects do have a TokenType field). C++ is statically typed and that means that type information is determined at compile-time and is not included in the generated code. Writing code in the parser that relies on some TokenTypes being *LexicalTokens* and other tokens just *Tokens* is a bad idea. What if we change either of the token classes in the future?

What we want is to be able to call delete to free the token regardless of which type it is. When we call delete on a Token object then nothing is done. When we call delete on a LexicalToken then the lexeme is deleted, too. We want the right method called depending on what the pointer is pointing to and not depending on the type of the pointer. This is *polymorphism* in action. To have the right method called depending on the type of an object the method must be polymorphic, which literally means many forms. A polymorphic method is actually two or more methods. The right one gets called depending on what object it is being called on.

The getLex method above is another instance of a polymorphic method. The right getLex will get called depending on the object it is called on. It is probably debatable whether the getLex method should be defined on the Token class at all. However, without it being defined there, we would have to first cast or coerce the Token object we get back from the scanner to a LexicalToken object so we could call getLex on it. This is avoided by defining getLex on the base class.

How does polymorphism work? For it to work the object must have control over which method is to be called because the pointer to the object doesn't tell the compiler anything. In fact, the decision can't be made until run-time. Polymorphism is a dynamic, or run-time, problem.

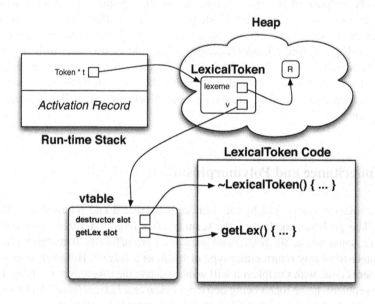

Fig. 3.3: Implementation of Polymorphism

Figure 3.3 shows a picture of the structure involved in making polymorphism work. While a particular C++ compiler might not follow this structure exactly, the principle would be the same regardless of compiler. Here are the steps that are taken to make polymorphism work.

1. An object like a LexicalToken is created in some code. In this example it's created in the scanner.
2. When the object is created, it's created as a LexicalToken. The type is known by the compiler at this point. The object is initialized by calling the LexicalToken constructor.
3. The compiler looks at the class description for LexicalToken and sees two virtual functions. It knows that virtual functions require a vtable to be created. The name vtable stands for virtual function table. At a given offset in every object there is a pointer to a vtable. Some objects may not have a vtable if they don't have any virtual functions, but the vtable pointer is still there.
4. The compiler finds the vtable associated with the class of the object being created. There is only one vtable for a class and the compiler knows where each vtable is stored. Code is generated to store the pointer to the vtable in the object.
5. When a virtual method like getLex or the destructor is called the compiler looks at the class and sees that the method is a virtual method. It knows that it must look up the code in the vtable.
6. The compiler generates code to look in the vtable at a specified offset depending on the virtual method being called. The destructor is always in the first slot, the getLex is always in the second slot regardless of whether it is a Token object, LexicalToken object, or some new object class that inherits from either. New entries can be added to the end of a vtable for new virtual functions in derived classes, but existing entries can NEVER be altered. The compiler must be able to count on the vtable structure remaining the same.
7. The vtable entry chosen contains a pointer to the correct code to be called. At run-time the program jumps to the correct method and polymorphism has just occurred.

Given that you understand the process described above, there are some things that should now be clear about polymorphism.

- A constructor can't be polymorphic, nor would we want it to be. When we construct an object we always specify the type of object we are creating. Polymorphism doesn't make sense for constructors.
- Once a method is declared polymorphic (i.e. virtual) it may NEVER become non-polymorphic again. Said another way, once it's in the vtable it stays in the vtable. That's why the destructor and getLex() method in LexicalToken aren't declared virtual, yet they still are. Once virtual, always virtual.
- It is permissable for a function to be declared non-virtual in a base class and then to become virtual in a sub-class since the function can be added to the vtable in a subclass without any problems. However, this is a bad idea in general. In the base class the function would not be polymorphic. You can try this out. Get the cppcalc code and change the definition of token.h so that Token's getLex method is not virtual (delete the keyword virtual) but leave the LexicalToken getLex method as virtual. It should compile but you won't get the right result. This is because when getLex is called on a Token pointer the non-polymorphic `Token::getLex()`

method will be called and not the polymorphically correct getLex. Be sure to *make clean* and then *make* between trial runs to get everything to recompile.

- Polymorphic method calls are less efficient than normal function calls. There is some overhead involved in making the function call. It amounts to two extra load instructions in most CPUs to complete the function call. Some hardware may be optimized to reduce the overhead of these loads. However, for most applications this is a small price to pay for code reuse.

3.5 A Historical Look at Parameter Passing

When the C programming language was designed passing parameters to functions was a fairly new concept. The Algol language designed during the 60's stressed the importance of structured programming. Passing parameters allowed code to be written in a modular way. The same code could be called from multiple points within a program without worrying about setting global variables to the right values first.

At the time that C was being developed, Niklaus Wirth was developing the Pascal programming language. He also had a goal of supporting modular programming. In the Pascal language, the programmer decided when writing a function or procedure which parameters should be modified and which shouldn't. Pascal programmers had to declare parameters to be value or variable parameters.

Example 3.13

Here is an example of a Pascal function with both value and variable parameters.

```
1   procedure lookup_term(name:ident_type; var found:boolean;
2                                          var place:integer);
3
4   var k:integer;
5
6   begin
7   found:=false;
8   for k:=1 to non_term_index do
9     if terminal[k].ident = name then
10      begin
11      found:=true;
12      place:=k;
13      end;
14  end;  (*PROCEDURE*)
```

In the code above the name parameter is a value parameter. That means a copy of the value is passed to the procedure lookup_term. By passing a copy of the value, the caller of the procedure can be assured that any variable passed as the first parameter will not be modified since it is a value parameter. To call the procedure you might write something like:

```
lookup_term(aName, found, aPlace);
```

The found and place parameters are declared as variable parameters. These parameters may be modified and this parameter passing method is often called pass by reference. That means that after calling the procedure, the original variables found and aPlace will be changed to the values assigned in the procedure. That also means the caller of the procedure must pass a variable into the function for the second and third parameters. Depending on the size of the ident_type type, the procedure may have a large value to copy as the first parameter. However, in the interest of good programming style, Niklaus Wirth ignored such efficiency issues and developed a programming language that enforced modular-structured programming.

Ritchie, when developing C was developing it for a different purpose. He wanted a programming language that supported modular programming, but was also efficient, convenient, and concise. In doing so he made some compromises. In C, there was only one parameter passing mechanism, pass by value.

Example 3.14

Consider the following program:

```
1   #include <stdio.h>
2
3   struct Point {
4      int x;
5      int y;
6   };
7
8   void testit(struct Point p) {
9      p.x = 0;
10     p.y = 0;
11  }
12
13  int main(int argc, char* argv[]) {
14     struct Point myPoint;
15     myPoint.x = 10;
16     myPoint.y = 10;
17     testit(myPoint);
18     printf("x = %d, y = %d\n",myPoint.x,myPoint.y);
19     return 0;
20  }
```

When compiled and run the program creates a structure called myPoint of type Point. Structures are the C equivalent of a record in Pascal and the equivalent of a class in C++ without any member functions and public access to all fields within the structure. The variable myPoint is passed by value and the copy of it is modified in the function called testit. However, the original value of myPoint is not changed. The program prints 10, 10 as the output. If a C programmer wishes to pass an argument by reference, he or she must pass a pointer to the original space.

☞ Practice 3.4

Considering what you learned about the run-time stack and calling functions in the first chapter, describe in detail what happens in the following program with regards to the run-time stack and the variables within the program.

```c
#include <stdio.h>

struct Point {
   int x;
   int y;
};

struct Point makePoint() {
   Point aPoint;
   aPoint.x = 0;
   aPoint.y = 0;
   return aPoint;
}

void testit(struct Point p) {
   p.x = 0;
   p.y = 0;
}

int main(int argc, char* argv[]) {
   struct Point myPoint = makePoint();
   myPoint.x = 10;
   myPoint.y = 10;
   testit(myPoint);
   printf("x = %d, y = %d\n",myPoint.x,myPoint.y);
   return 0;
}
```

Example 3.15

To pass the myPoint by reference in C the following code would need to be written.

```c
#include <stdio.h>

struct Point {
   int x;
   int y;
};

void testit(struct Point* p) {
   p->x = 0;
   p->y = 0;
}

int main(int argc, char* argv[]) {
```

```
14    struct Point myPoint;
15    myPoint.x = 10;
16    myPoint.y = 10;
17    testit(&myPoint);
18    printf("x = %d, y = %d\n",myPoint.x,myPoint.y);
19    return 0;
20  }
```

In the version of the code in example 3.15 the variable p is a pointer to a Point structure. By passing the pointer to the original structure the original structure can be updated. The output from this version of the program is 0, 0. There are some things to take note of in this version of the program.

First, in the testit function the code that was p.x = 0 is now p->x = 0. Since p is now a pointer, we must use pointer notation to dereference p. In Java and many other languages we can dereference a pointer or reference by using the dot (i.e. a period) notation, but not in C. In C, to dereference a pointer you must either write an arrow (i.e. ->) or use an asterisk (i.e. *). So, equivalently we could write (*p).x = 0 to set the x field of the Point pointed to by p to 0.

The second and more important difference is that the caller of the testit function must now realize that the parameter is a pointer and must pass the address of the Point to the function. That is why &myPoint is written. The &, called an ampersand, is the addressof operator. It passes the address of myPoint to the testit function. Requiring the caller of a function to remember to write the address of operator for reference parameters is probably not a good idea to promote modular programming. Now, the programmer who writes the function does not maintain entire control over how parameters are passed. The caller of the function must also remember to call it in the right way.

☞ Practice 3.5

What could go wrong in the following version of the code? Carefully trace the code by hand to see what the mistake is here.

```
1   #include <stdio.h>
2
3   struct Point {
4     int x;
5     int y;
6   };
7
8   struct Point* makePoint() {
9     Point aPoint;
10    aPoint.x = 0;
11    aPoint.y = 0;
12    return &aPoint;
13  }
14
15  void testit(struct Point* p) {
16    p->x = 0;
17    p->y = 0;
```

```
18    }
19
20    int main(int argc, char* argv[]) {
21      struct Point* myPoint = makePoint();
22      myPoint->x = 10;
23      myPoint->y = 10;
24      testit(myPoint);
25      printf("x = %d,  y = %d\n",myPoint->x,myPoint->y);
26      return 0;
27    }
```

Around 1980, Bjarne Stroustrup was working on the C++ programming language. This idea of passing parameters by reference by passing in a pointer to the data had long been a criticism of the C language. So, when Dr. Stroustrup designed C++ he included the idea of passing parameters by reference.

Example 3.16

The code in this example passes the Point data by reference.

```
1    #include <stdio.h>
2
3    class Point {
4    public:
5      int x;
6      int y;
7    };
8
9    void testit(Point& p) {
10      p.x = 0;
11      p.y = 0;
12    }
13
14    int main(int argc, char* argv[]) {
15      Point myPoint;
16      myPoint.x = 10;
17      myPoint.y = 10;
18      testit(myPoint);
19      printf("x = %d,  y = %d\n",myPoint.x,myPoint.y);
20      return 0;
21    }
```

It may be somewhat confusing to the person just learning C++. Pay close attention if you are one of those people. First, the struct Point is replaced with a class Point since C++ supports classes. All fields within the class are declared public in this example, generally a bad thing to do. Then, in this example, the ampersand moves from the caller of the function to the function definition. When a formal parameter of a function in C++ has an ampersand after the type it means the parameter is passed by reference. A reference is like a pointer in C, but the caller of the function does not need to pass the address of the data to the function. C++ does this under the covers. In addition, when a parameter is declared as a reference parameter, C++

also dereferences the reference automatically when it is evaluated. So, instead of writing p->x = 0 we may once again write p.x = 0 which is more consistent with other modern programming languages.

So, there are two parameter passing mechanisms in C++, but because pointers may be passed as parameters there are for all practical purposes three ways of passing parameters in C++. Pass by value copies the value, pass by pointer reference passes a copy of the pointer but the pointer and its copy both point to the original data, and pass by reference.

In contrast to C++, the Java programming language passes built-in types like int, double, float, and char by value. All objects in Java are passed by reference. Java is less flexible in its parameter passing, but this simplifies many aspects of the language.

In fact, passing objects by value leads to a lot of complexity in C++. Because objects can and regularly do have pointers to other objects within them, making a copy of an object when it is passed by value is too complicated for C++ to do by itself. Programmers must write special a special method as part of the class declaration to make pass by value work correctly. The special pass by value method is called a copy constructor. A copy constructor makes a copy of the object in the way dictated by the programmer. It would be the equivalent of the clone method in Java. Copy constructors are not discussed here.

3.6 Const in C++

The keyword *const* in C++ can be used in a variety of situations. Perhaps the simplest situation is where you want to declare a constant. You can, for instance, write

```
const int maxVal = 100;
```

to declare a constant value maxVal. Declaring constants can save time when changing the value of a constant. They can also help you avoid introducing errors when changing constant values by requiring you to change fewer lines of code. More importantly, they can help make your code self-documenting.

The *const* keyword can also be used in other circumstances in C++. Perhaps the most important use of *const* is to solve the problem of pass by value copying large objects in C++. The problem is that objects, when passed by value, are copied. Passing by reference would seem an obvious solution to this and it is why objects are passed by reference in languages like Java. Pass by reference is more efficient. However, passing a parameter by reference means that the called function or method may alter the parameter. Languages like Java ignore this possibility. It is up to the programmer to know whether an object is altered by a function or not. Usually, the programmers relies on documentation in Java to tell them whether or not the parameter is modified.

In C++, it is possible to declare that a parameter is not modified by a method by either declaring the parameter is passed by value or by declaring the parameter is passed by constant reference. Declaring it as a constant reference means that we wish to pass the parameter efficiently (without making a copy) but we promise not to alter the values of any fields within the object.

Example 3.17

In this code, the Point example has been rewritten to call a method called `printPoint` to print the point data. However, the `printPoint` method will not modify the point so the parameter is declared to be a constant reference to a point.

```
1   #include <stdio.h>
2
3   class Point {
4   public:
5     Point(int x, int y);
6     int getX() const;
7     int getY() const;
8     void setX(int x);
9     void setY(int y);
10
11  private:
12    int x;
13    int y;
14
15  };
16
17  Point::Point(int x, int y) {
18    this->x = x;
19    this->y = y;
20  }
21
22  int Point::getX() const {
23    return x;
24  }
25
26  int Point::getY() const {
27    return y;
28  }
29
30  void Point::setX(int x) {
31    this->x = x;
32  }
33
34  void Point::setY(int y) {
35    this->y = y;
36  }
37
38  void testit(Point& p) {
39    p.setX(0);
40    p.setY(0);
41  }
42
```

```
43    void printPoint(const Point & p) {
44      printf("Point(%d,%d)\n",p.getX(),p.getY());
45    }
46
47    int main(int argc, char* argv[]) {
48      Point myPoint(10,10);
49      testit(myPoint);
50      printPoint(myPoint);
51      return 0;
52    }
```

In the example above you may notice that the methods getX() and getY() of the Point class are declared as const methods. This is an example of how *constness* may creep into your program. When the printPoint method was declared to take a constant reference to a Point object, it meant that printPoint could not call any methods on the Point class that could potentially modify the Point object. That meant that the getX() and getY() methods now need to be declared to be const themselves. When const appears after a method declaration it means that the method is an accessor method of the class and will not modify the object data in any way. Once you start using constant references to pass parameters efficiently, it will creep into your program and require other parts of your program to be declared const as well. This isn't necessarily bad. When used correctly it can help programmers avoid illegal memory references and other subtle problems.

Notice the getType, getCol, getLine, and getLex member functions of Token are all declared as const member functions. These functions are accessor functions of the Token class and therefore may be declared as const functions. This might be useful if the programmer wishes to write some code that passes Token objects as constant references at some point.

One interesting nuance of this discussion is worth taking note of. If a class contains all const member functions, if there are no mutator methods on the class, then whether an object of that class is passed by value or passed by reference becomes a moot point. The idea of a *constant* reference can go away since all objects of that class must be constant references. This would certainly simplify matters.

In addition, if a language were to only support immutable objects, then the idea of pass by value and pass by reference could go away entirely along with all the complexity of the parameter passing system. Immutable objects are objects that once created cannot be altered. If you want to modify an object, you create a new object with the modified fields instead. Standard ML is one such language. In fact, many functional languages only support immutable data.

One other way to eliminate the complexity of parameter passing is to support only pass by value or pass by reference. Java only supports pass by value of primitive types because they are small in size and may be passed by value efficiently. It only supports pass by reference for objects as stated earlier. This simplifies the language support for parameter passing and presents a clear and consistent model for the programmer. Java programmers know that all objects are passed by reference.

3.7 The AST Classes

According to the diagram in figure 3.1 the parser must build an abstract syntax tree of the expression to be evaluated. To do this a collection of classes must be declared. The AST classes can be designed as a hierarchy of classes where each type of AST class represents one type of node in an abstract syntax tree.

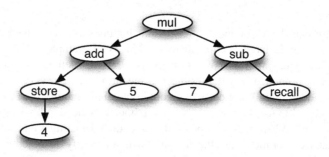

Fig. 3.4: Abstract Syntax Tree of expression in figure 3.1

Consider the abstract syntax tree in figure 3.4. There are mul, add, sub, store, recall, and number nodes within the tree. If we abstract away from the details a bit we also see there are some nodes which have no children, some with one child, and some with two children.

The tree could be evaluated to produce the value of the expression. A postfix traversal of the tree would yield the value. For instance consider the steps in the postfix traversal given here.

1. The traversal begins by recursively descending the left side of the tree down to the 4 node. Visiting that node returns 4.
2. The store node takes the 4 and stores it in the calculator's memory. It also returns the 4.
3. The add can't be visited yet since it has a right child (the 5). The 5 node is visited and returns the 5.
4. The add node can now be visited. It takes the 4 and the 5, adds them together and returns the 9.
5. The mul node can't be visited until its right child is visited. Postorder traversal of the sub node calls the traversal on the 7 node, which returns 7.
6. The sub node still can't be visited yet. The recall node is traversed and returns the value in the calculator memory, the 4.
7. The 7-4 is computed by visiting the sub node and returns 3.
8. Visiting the mul node computes 9*3 or 27.

This evaluate procedure can be accomplished by writing a polymorphic *evaluate* method for an abstract syntax tree. First, we'll define an abstract base class for the AST. We'll call this class AST. An abstract base class never has any objects of its type created. Its purpose is to serve as a base class for other classes. The AST class is abstract because the two virtual functions have a *0* representation. However, the destructor must still be written because it must have a slot in the vtable.

Because an abstract syntax tree consists of null-ary, unary, and binary nodes, the AddNode, SubNode, MulNode, DivNode, StoreNode, NumNode, and RecallNode classes can inherit from one of the three classes below. Each class will implement its own destructor and evaluate method. The polymorphic evaluate methods will implement the postorder traversal of the tree described above.

Example 3.18

Here is the AST header file containing three AST class declarations.

```
1  #ifndef ast_h
2  #define ast_h
3
4  using namespace std;
5
6  class AST {
7    public:
8      AST();
9      virtual ~AST() = 0;
10     virtual int evaluate() = 0;
11   };
12
13   class BinaryNode : public AST {
14     public:
15       BinaryNode(AST* left, AST* right);
16       ~BinaryNode();
17       AST* getLeftSubTree() const;
18       AST* getRightSubTree() const;
19     private:
20       AST* leftTree;
21       AST* rightTree;
22   };
23
24   class UnaryNode : public AST {
25     public:
26       UnaryNode(AST* sub);
27       ~UnaryNode();
28       AST* getSubTree() const;
29     private:
30       AST* subTree;
31   };
32
33   #endif
```

Example 3.19

The implementation of three AST classes is provided here.

```
1  #include "ast.h"
2  #include <iostream>
3  #include "calculator.h"
4
5  //uncomment the next line to see the destructor calls
6  //#define debug
7
8  AST::AST() {}
9
10 AST::~AST() {}
11
12 BinaryNode::BinaryNode(AST* left, AST* right):
13    AST(),  leftTree(left),  rightTree(right)
14 {}
15
16 BinaryNode::~BinaryNode() {
17 #ifdef debug
18    cout << "In BinaryNode destructor" << endl;
19 #endif
20    try {
21       delete leftTree;
22    } catch (...) {}
23    try {
24       delete rightTree;
25    } catch (...) {}
26 }
27
28 AST* BinaryNode::getLeftSubTree() const {
29    return leftTree;
30 }
31
32 AST* BinaryNode::getRightSubTree() const {
33    return rightTree;
34 }
35
36 UnaryNode::UnaryNode(AST* sub):
37    AST(),  subTree(sub)
38 {}
39
40 UnaryNode::~UnaryNode() {
41 #ifdef debug
42    cout << "In UnaryNode destructor" << endl;
43 #endif
44    try {
45       delete subTree;
46    } catch (...) {}
47 }
```

The destructors are provided above. The evaluate methods are not. The UnaryNode and BinaryNode classes inherit the abstract definition of evaluate from AST.

AddNode, SubNode, MulNode, DivNode, NumNode, StoreNode, and RecallNode must be derived from the appropriate class and the evaluate method must be overridden to implement the correct function.

☞ Practice 3.6

Write the declaration of the AddNode class.

☞ Practice 3.7

Write the implementation of the AddNode class. When writing the evaluate method for the AddNode class don't worry about how you get the right values. Just assume that those values are available if you call evaluate on the right object or objects. How it's done is unimportant in this exercise.

The implementation of the SubNode, MulNode, DivNode, NumNode, StoreNode, and RecallNode classes is left as an exercise for the reader.

3.8 The Scanner

Referring back to figure 3.1 the scanner reads characters from the input and builds Token objects that are used by the parser. To accomplish this, the scanner needs to read characters from a stream and decide how to group them into tokens. The parser will get tokens from the scanner by calling a getToken method. Sometimes the parser gets a token and needs to put it back. In that case a putBackToken method will put back the last token that was returned by getToken.

Example 3.20

Here is the declaration of the scanner in the header file scanner.h.

```
1  #ifndef scanner_h
2  #define scanner_h
3
4  #include <iostream>
5  #include "token.h"
6
7  class Scanner {
8  public:
9     Scanner(istream* in);
10    ~Scanner();
11
12    Token* getToken();
13    void putBackToken();
14
15  private:
16    Scanner();
```

```
17
18      istream* inStream;
19      int lineCount;
20      int colCount;
21
22      bool needToken;
23      Token* lastToken;
24   };
25
26   #endif
```

In line 16 above the default constructor for the Scanner class is declared private. By making the default constructor private a Scanner object can never be constructed that way. A Scanner should always be constructed over an input stream, so this is what we want.

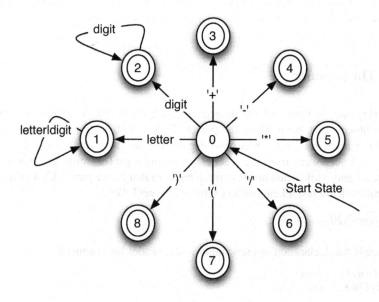

Fig. 3.5: A Finite State Machine for the Scanner

Internally, the scanner object is a finite state machine. A finite state machine (fsm) consists of a set of states and a set of transitions from one state to another based on the current character in the input. Figure 3.5 defines the Scanner's finite state machine. The fsm starts in state zero, reads one character and transitions to one of the eight states depending on the character. In state one there is a transition that stays in state one as long as a letter or digit is read. The fsm continues to read characters and make transtions until a character appears that has no transition from the current state. At that point, if the state is an accepting state (i.e. a double circle state) the string of characters is recognized as a token.

This fsm reads a token and returns it. The eof and keyword tokens are handled as special cases. Eof is handled in state zero when the end of stream is reached. State one compares the identifier to a list of keywords to see if it should return a keyword or identifier token. The complete implementation can be found in appendix A.

Whitespace is read and thrown away by the scanner. Whitespace consists of blanks, tabs, and newline characters. When a character is read that is not recognized by a transition from the current state, the fsm returns the current token (since states 1-8 are all accepting states) saving the current character for later. The next call to getToken resumes with the current character. If there is no valid transition from state zero on the current character the fsm returns the character as an unrecognized token.

An fsm is a model of computation for recognizing strings of characters. An fsm is easily implemented using a while loop, a switch statement, and one variable to record the current state. Fsm's are used in many contexts including network protocol implementations, pattern recognition, simulations, and of course language implementation. There are tools to build powerful fsm's. However, it's good to see a hand-written one to aid in understanding some of the theory behind fsm's, too.

3.9 The Parser

Figure 3.1 shows the parser reading tokens and producing an abstract syntax tree as its output. In section 2.7 on page 31 a parser was defined as a program that given a sentence (i.e. a string of tokens), checks to see if the sentence is in the language of a given grammar. The parser that is discussed in this section is a top-down parser. The parser will build the AST from the top-down. In reality, top-down is a bit of a misnomer. While the construction of the tree starts at the top, the tree is actually built bottom-up by a recursive descent of the tree. That's why top-down parsers are also called recursive descent parsers.

To begin to design a parser there must be a grammar to model it after.

Example 3.21

This is the Calculator language's grammar.

$Prog \rightarrow Expr\ EOF$
$Expr \rightarrow Expr + Term \mid Expr - Term \mid Term$
$Term \rightarrow Term * Storable \mid Term / Storable \mid Storable$
$Storable \rightarrow Factor\ S \mid Factor$
$Factor \rightarrow number \mid R \mid (Expr)$

A recursive descent parser is, believe it or not, recursive. The implementation of the parser is given to us by its grammar. In the implementation, each nonterminal becomes a function in the parser. Each rule in the grammar is part of a function that is named by the nonterminal on the left side of the arrow in the rule. In the grammar above each line would correspond to a function in the parser. Each appearance of a nonterminal on the right hand side of a production is a function call. Each appearance of a token on the right hand side of a production is a call to the scanner to get a token. From this definition, writing the parser is pretty straightforward.

A First Attempt at Writing the Parser

The parser will read the tokens and build an abstract syntax tree like the one figure 3.4. To write the top-down parser of these expressions each nonterminal becomes a function. The grammar dictates how to write the parser. The body of each function is given by the right hand side of its corresponding production.

Example 3.22

The Prog and Expr functions for the Parser

```
1   AST* Parser::Prog() {
2       AST* result = Expr();
3       Token* t = scan->getToken();
4
5       if (t->getType() != eof) {
6           cout << "Syntax Error: Expected EOF, found token "
7                << " at column " << t->getCol() << endl;
8           throw ParseError;
9       }
10
11      return result;
12  }
13
14  AST* Parser::Expr() {
15      AST* e = Expr();
16      Token* t = scan->getToken();
17      ...
18  }
```

There is a big problem with the Expr function given above. It is recursive and there is no base case. This means if you call the Expr function, it will go into infinite recursion resulting in run-time stack overflow. The grammar given above isn't suited for top-down parsing.

A Better Attempt at Writing a Top-Down Parser

The problem in the previous section is that the grammar is not LL(1). For a grammar to be LL(1) means that the choice of which production to apply next in a left-most derivation of a sentence can be made by looking ahead at the next token. The number one in LL(1) means that only one token of lookahead is needed to decide which production to use. Although the grammar above is LALR(1), it is not appropriate for constructing a recursive descent parser. An LL(1) grammar is needed to build a recursive descent or top-down parser. An LALR(1) grammar is a grammar that can be given to a program to construct a reverse right-most derivation of a sentence in the grammar looking ahead at only the next token in the sentence. This is what a bottom-up parser does and bottom-up parser generators can take a grammar like the one above and automatically construct a parser for it. Because bottom-up parsers are harder to write, we usually rely on a parser generator program to write the parser for us when generating a bottom-up parser.

Top-down parsers are much simpler to write and are typically written by hand. However, to create a top-down parser you have to have an LL(1) grammar. Fortunately, it is relatively easy to convert an LALR(1) grammar to an LL(1) grammar. There are two steps involved.

1. Eliminate left recursion.
2. Perform left factorization where appropriate.

Eliminate Left Recursion

Eliminating left recursion means eliminating rules like $Expr \rightarrow Expr + Term$. Rules like this are left recursive because the *Expr* function would first call the *Expr* function in a recursive descent parser as in example 3.22 above. Without a base case first, we are stuck in infinite recursion (a bad thing). To eliminate left recursion we look to see what Expr can be rewritten as when deriving a sentence. In this case, Expr can only be replaced by a Term so we replace Expr with Term in the productions. Then, we add a new nonterminal to represent the rest of the production from the LALR(1) grammar. In this case, the + *Term* and the - *Term* are left after we replace the initial *Expr* in the productions in the grammar above. The usual way to eliminate left recursion is to introduce a new nonterminal to handle all but the left recursive nonterminal. Two rules in the grammar are left recursive and must be rewritten.

Example 3.23

An LL(1) Calculator Language Grammar

> *Prog → Expr EOF*
> *Expr → Term RestExpr*
> *RestExpr → + Term RestExpr | − Term RestExpr | <null>*
> *Term → Storable RestTerm*
> *RestTerm → * Storable RestTerm | / Storable RestTerm | <null>*
> *Storable → Factor S | Factor*
> *Factor → number | R | (Expr)*

In this example the *Expr → Expr + Term | Expr − Term | Term* is replaced by the second and third lines of the grammar given above. Likewise, the left recursion in *Term → Term ∗ Storable | Term/Storable | Storable* is rewritten as the fourth and fifth lines of the grammar above.

Perform Left Factorization

Left factorization isn't needed on this grammar so this step is skipped. Left factorization is needed when the first part of two or more productions are the same and the rest of the similar productions are different. Left factorization is important in languages like Prolog because without it the parser may have to backtrack. Since backtracking won't work when reading something from an input stream you must perform left factorization by writing a new rule that handles the common prefix of the two offending rules. However, it isn't needed in C++ if you recognize the common prefix and code the function appropriately.

Translating the LL(1) Grammar to C++

Once you have an LL(1) grammar you use it to build a parser as follows. The following construction causes the parser to return an abstract syntax tree for the sentence being parsed.

1. Construct a function for each nonterminal. Each of these functions should return a node in the abstract syntax tree.
2. Depending on your grammar, some nonterminal functions may require an input parameter of an abstract syntax tree (ast) to be able to complete a partial ast that is recognized by the nonterminal function.
3. Each nonterminal function should call getToken on the scanner to get the next token as needed. If after getting the token, the code determines it didn't need the token after all, the nonterminal function should call the scanner's putBackToken function to put the token back. If the parser is based on an LL(1) grammar, it should never have to put back more than one token at a time.

4. The body of each nonterminal function is a series of if statements that choose which production to expand upon depending on the value of the next token. The body of the function is determined by the productions of the grammar with the given nonterminal on the left hand side of the arrow.

Example 3.24

This is the Parser's header file, "parser.h".

```
1   #ifndef parser_h
2   #define parser_h
3
4   #include "ast.h"
5   #include "scanner.h"
6
7   class Parser {
8     public:
9       Parser(istream* in);
10      ~Parser();
11
12      AST* parse();
13
14    private:
15      AST* Prog();
16      AST* Expr();
17      AST* RestExpr(AST* e);
18      AST* Term();
19      AST* RestTerm(AST* t);
20      AST* Storable();
21      AST* Factor();
22
23      Scanner* scan;
24  };
25  #endif
```

The construction described here leads to a class declaration for the Parser class as given in example 3.24. The only public function of the parser is the parse function which returns an AST of the expression that was parsed. The private functions correspond to the nonterminals of the grammar. These functions are private because the user of a parser doesn't need to know the details of how the parser works. You do, however!

The construction above is very simple, but can be confusing without an example. Consider the LL(1) grammar given above. Assume that you have two classes called AddNode and SubNode that are derived from the BinaryNode class.

Example 3.25

The Parser's Prog and Expr Functions

```
1   AST* Parser::Prog() {
2       AST* result = Expr();
3       Token* t = scan->getToken();
4
5       if (t->getType() != eof) {
6           cout << "Syntax Error: Expected EOF, found token "
7                << "at column " << t->getCol() << endl;
8           throw ParseError;
9       }
10
11      return result;
12  }
13
14  AST* Parser::Expr() {
15      return RestExpr(Term());
16  }
```

The code in example 3.25 throws an exception if an error is discovered during parsing. You would normally take appropriate action during error conditions, but throwing an exception is a legitimate way to deal with a parsing problem. The Prog function returns a reference to an AST, which is the abstract syntax tree representing the expression that was parsed.

The Expr function in lines 14-16 corresponds to the *Expr* rules in the grammar in example 3.23. The rule says to first call the Term function. The result of calling this function is an AST (as all nonterminal functions return an AST).

☞ Practice 3.8

The RestExpr function is slightly different from the Prog and Expr functions. The RestExpr function has an AST parameter which we'll call e. The RestExpr function first gets a token and then decides what to do based on that token. If it is an *add* token it builds a new AST AddNode with the part of the tree given to it (i.e. e) as the left subtree and the result of calling Term as the right subtree. The subtract AST nodes are handled similarly. Otherwise, there wasn't a token that the RestExpr knows about so the token is put back and the AST e is returned as its AST.

Write the RestExpr function described here. Remember you can refer to the grammar in example 3.23.

The remainder of the parser implementation can be patterned after the code in example 3.25 using the grammar in example 3.23 as a guide. The remainder of the code is left as an exercise.

3.10 Putting It All Together

One more class is required to tie together the pieces that have been developed in this chapter. The Calculator class contains a memory location that can hold a stored value. The value can also be retrieved on demand. The calculator can evaluate an expression that is given to it as a string.

Example 3.26

The Calculator's header file, "calculator.h"

```
1   #ifndef calculator_h
2   #define calculator_h
3
4   #include <string>
5
6   using namespace std;
7
8   class Calculator {
9    public:
10     Calculator();
11
12     int eval(string expr);
13     void store(int val);
14     int recall();
15
16    private:
17     int memory;
18   };
19
20   extern Calculator* calc;
21
22   #endif
```

The extern statement above means that someplace in the project a calc pointer will be declared as a global variable. This global variable is needed in the abstract syntax tree implementation so the AST can have access to the calculator's memory location. By declaring calc extern we tell the compiler that calc is a global variable that can be accessed from any module that includes this declaration. Global variables are generally a bad idea. This is one case where it is justified. Without it just about every object presented in this chapter would have to keep a reference to the calculator. That's a lot of overhead to have access to one little memory location.

Example 3.27

The Calculator class implementation

```
1   #include "calculator.h"
2   #include "parser.h"
3   #include "ast.h"
4   #include <string>
5   #include <iostream>
6   #include <sstream>
7
8   Calculator::Calculator():
9       memory(0)
10  {}
11
12  int Calculator::eval(string expr) {
13
14      Parser* parser = new Parser(new istringstream(expr));
15
16      AST* tree = parser->parse();
17
18      int result = tree->evaluate();
19
20      delete tree;
21      delete parser;
22
23      return result;
24  }
25
26  void Calculator::store(int val) {
27      memory = val;
28  }
29
30  int Calculator::recall() {
31      return memory;
32  }
```

The Calculator evaluates an expression by creating an istringstream over the string containing the expression. An istringstream is an input stream constructed from a string. This stream is passed to the Parser constructor. The Parser in turn constructs a Scanner object over the stream to get the tokens from the string.

If all goes well, the string is parsed and the parser returns an AST. The tree is then evaluated (polymorphically) to yield the result. The result is returned to the main program to be printed. This flow of data is depicted in the dataflow diagram shown in figure 3.1.

Example 3.28

The main function from "calc.C"

```
1   #include <iostream>
2   #include <sstream>
3   #include <string>
4   #include "calcex.h"
5   #include "calculator.h"
6   using namespace std;
7
8   Calculator* calc;
9
10  int main(int argc, char* argv[]) {
11      string line;
12
13      try {
14
15          cout << "Please enter a calculator expression: ";
16
17          getline(cin, line);
18
19          calc = new Calculator();
20
21          int result = calc->eval(line);
22
23          cout << "The result is " << result << endl;
24
25          delete calc;
26
27      }
28      catch (Exception ex) {
29          cout << "Program Aborted due to exception!" << endl;
30      }
31  }
```

The global variable is declared in this last module. The main function gets the input line from the user, creates the calculator, and calls eval on the calculator giving it the input line. The result is printed to the screen.

3.11 Exercises

1. What's the value of (R+7)/4S if the memory contained 4 prior to evaluating this expression?
2. What is the value of the memory location after evaluating the previous expression?
3. What does the abstract syntax tree look like for the expression (R+7)/4S?
4. How could the calculator language be modified to allow more than one memory location like modern calculators? Discuss what changes would be required to implement this enhanced calculator language.
5. Complete the calculator interpreter by downloading the code given in this chapter and finishing the implementation in parser.C, ast.C, and ast.h. The rest of the project is provided.

   When you download the code you will want to unzip the package with some sort of unzip program. On Linux you can issue the command,

   ```
   unzip cppcalc.zip
   ```

   Then you can make the program and run it. Here is an example of making and running you can use to get started.

```
1   $ unzip cppcalc.zip
2   $ cd cppcalc
3   $ make
4   make: `calc' is up to date.
5   $ make clean
6   rm -f *.o
7   rm -f calc
8   $ make
9   g++ -g -c calc.C
10  g++ -g -c scanner.C
11  g++ -g -c token.C
12  g++ -g -c ast.C
13  g++ -g -c parser.C
14  g++ -g -c calculator.C
15  g++ -g -o calc calc.o scanner.o token.o ast.o parser.o \
16      calculator.o
17  $ calc
18  Please enter a calculator expression: 5 + 4
19  The result is 9
20  $
```

Commands that you enter are preceded by a dollar sign. The *make clean* above tells make to use the *clean* rule to erase all compiled files and make the project from scratch. See the file *Makefile* for the clean rule or look in the chapter at the section on the make utility.

The program will compile and add two integers together as provided. Your job is to extend the project to the full calculator language. This requires changes to the parser.C and ast.C modules. The parser.C changes are highlighted in section 3.9.

You can complete the parser by completing the functions that are incomplete in the parser.C file. These functions can be patterned after the code presented in the chapter.

The parser code will require that you build AST nodes for storing values and for recalling values from the calculator's memory. You will also need multiply and divide nodes in the abstract syntax tree. These new node types can be added to the ast.C and ast.h files using the existing code as a pattern.

The store and recall nodes in the AST will need to access the memory location of the calculator. The global variable called *calc* can be used to access the calculator's memory. The line

```
calc->store(6)
```

will store 6 in the calculator's memory. Similarly, the expression *calc->recall()* will retrieve the value stored in the memory of the calculator.

6. Once you have completed the project described above extend the calculator language to allow more than one memory location to hold a value.

7. Modify the project to be a compiler instead of an interpreter. Instead of evaluating the expression, generate EWE code for it instead.

In addition, to make this interesting, add a new keyword to the language, called I, that when executed waits for user input before proceeding. The value returned by the call to I is the value entered at the keyboard.

This project can be implemented with a few modifications. First, the evaluate function of the abstract syntax tree will print code to a file called "a.ewe" instead of directly evaluating the expression. To print to a file in C++ you create an ofstream object. The changes can be made in several places but it will work if you create the ofstream in the main function and pass it to the constructor of the Calculator object. The object can be passed as an ostream which ofstream inherits from. Pass the ofstream by reference to try out references in C++. Declare the ostream as a reference in your Calculator object as well. For ofstream to be defined you must add the <fstream> include statement in your code. The Calculator object should have an additional method to return the ostream reference when asked so the evaluate functions in your abstract syntax tree can get the ofstream when printing code. The code that should be printed is EWE code. You will want to employ a correct model of computation when generating EWE code so you can systematically generate the required code for the expression.

To write to an ostream is just like writing to cout. See the cout write statements in calc.C for examples of how this is done.

8. C++ allows parameters to be passed by value, reference, or pointer. Modify the calculator program to pass Tokens by reference instead of by pointer.

9. Modify the calculator program to pass Tokens by value instead of by pointer.

10. Start with a fresh copy of the C++ calculator code. See exercise 5 for directions on downloading and unzipping the files. Compile it with the *make clean* and *make* commands. Run the program to verify that it does correctly add two integers together.

    Then change the *token.h* file and remove *virtual* keyword from the *getLex* method of the*Token* class but leave the *LexicalToken* . Run the program again and it will likely not add two integers together correctly. Explain why this happens. What happened to the C++ code that removing the keyword

3.12   Solutions to Practice Problems

These are solutions to the practice problems . You should only consult these answers after you have tried each of them for yourself first. Practice problems are meant to help reinforce the material you have just read so make use of them.

Solution to Practice Problem 3.1

1. 81
2. 6
3. Depends on the initial value of memory. It could be an error. Assuming the calculator starts with 0 in memory the answer would be 3. If the calculator is written to evaluate more than one expression in a session then the memory might contain the last value stored.
4. 10

Solution to Practice Problem 3.2

Below are to commands the make file would execute. The Makefile should probably have included a dependency of parser.C on the ast.h as well. But, if you get into trouble, type `make clean` to start over.

```
1  g++ -g -c ast.C
2  g++ -g -c calculator.C
3  g++ -g -o calc calc.o scanner.o token.o ast.o parser.o
4          calculator.o
```

Solution to Practice Problem 3.3

1. There are number, keyword, add, keyword tokens in this one.
2. The tokens are: lparen, number, add, number, rparen, keyword, times, keyword.

Solution to Practice Problem 3.4

The program uses pass by value so the Point data is copied between each of the function calls.

1. An activation record is pushed on the stack for the main function containing the variabie called `myPoint`.

2. Immediately the `makePoint` method is called pushing a new activation record on the stack. The new activation record has its own copy of a point called `aPoint`. The `aPoint` variable is initialized to (0,0) and then when `aPoint` is returned the activation record is popped from the run-time stack and the data is copied back into the `myPoint` variable.
3. The `myPoint` variable is changed to (10,10).
4. When `testit` is called another copy of a point is made in the run-time stack's new activation for the call to `testit` and that variabie is initialized to (0,0). This does not change the value of the `myPoint` variable.
5. Finally, the activation record for the call to `testit` is popped returning to main to finish the program by printing the (10,10) to the screen.

Solution to Practice Problem 3.5

In this example, the `makePoint` function returns a pointer to the `aPoint` variable. However that variable lies within the `makePoint` activation record. That is generally a bad idea. When `makePoint` returns, the main function will now have a pointer to memory on the run-time stack that may be reused in the not too distant future. This will almost certainly be true once the program calls the `testit` function. The outcome of this program is not well-defined. It may work and it may not. The outcome of the program depends on the underlying architecture of the target platform and the code generated by the compiler.

This is precisely the reason Java does not contain an `addressof` operator. Problems like this occur all the time in C++ when inexperienced C++ programmers start writing code. It is a bad idea to get the address of a variable, yet this happens all the time in C++, especially when using arrays. In Java, the only address ever provided is when a programmer uses the `new` keyword to create an object. However, the address returned by new is a reference, which itself is not a pointer and is safer to use than a pointer. A reference must point to an object and may not be a pointer to just anywhere in memory. In addition, the only references in Java are references into the heap. No pointers into the run-time stack are allowed in the language or its underlying implementation.

Solution to Practice Problem 3.6

Here is the AddNode declaration.

```
1   class AddNode : public BinaryNode {
2   public:
3       AddNode(AST* left, AST* right);
4
5       int evaluate();
6   };
```

Solution to Practice Problem 3.7

Here is the AddNode implementation.

```
1   AddNode::AddNode(AST* left, AST* right):
2       BinaryNode(left,right)
3   {}
4
5   int AddNode::evaluate() {
6       return getLeftSubTree()->evaluate() +
7               getRightSubTree()->evaluate();
8   }
```

Solution to Practice Problem 3.8

Here is the RestExpr solution.

```
1   AST* Parser::RestExpr(AST* e) {
2       Token* t = scan->getToken();
3
4       if (t->getType() == add) {
5           return RestExpr(new AddNode(e,Term()));
6       }
7
8       if (t->getType() == sub)
9           return RestExpr(new SubNode(e,Term()));
10
11      scan->putBackToken();
12
13      return e;
14  }
```

3.13 Additional Reading

Time and space constraints prevent further discussion of C++. There are many interesting and important aspects of C++ that have not been covered here. These include, but are not limited to:

- Namespaces
- Copy constructors
- Type conversion operators
- Streams and stream operators
- Operator overloading

An excellent reference on the whole language is given in [11]. There are also plenty of C++ programming guides available both online and in book form. If you know how to program in Java another excellent reference is Timothy Budd's book, *C++ for Java Programmers*[5].

Chapter 4

Object-Oriented Programming with Ruby

This chapter introduces you to the Ruby language through the development of an interpreter for the calculator language like the C++ version presented in chapter 3. Unlike C++, Ruby is an interpreted language. All data in Ruby are objects which is different than C++ and Java. Ruby is also dynamically typed. This also is unlike C++ and Java. C++ and Java are object-oriented languages, but are statically typed. C++ is also a compiled language, not interpreted. While there are many differences between the two languages, there are also many similarities. Using Ruby and C++ to develop the same project can give you some insight into the differences and similarities of the two languages. This chapter will point out those similarities and differences.

Ruby was influenced by several older (relative to Ruby anyway) programming languages including Smalltalk and Perl. Being object-oriented, Ruby allows the programmer to declare his or her own classes. A class is a way of describing a set of objects with similar attributes. A class also contains the code to manipulate objects of the class.. The attributes of an object are called *instance variables*. The manipulation of objects is through special functions called *methods*. Every method in Ruby is a function and returns a value. This is the way Smalltalk was designed as well so Ruby gets this idea of methods returning values from Smalltalk.

Ruby is an interpreted language with garbage collection. It doesn't have the problem of memory leaks like C++ programs because you don't have to free memory yourself. Ruby is not a hybrid language implementation like Java. In other words, it is not compiled to an intermediate form and then interpreted. It is truly interpreted from an abstract representation of the source program. As a result, Ruby programs can run significantly slower than their C++ counterparts. Some work is being done with Ruby to transform it into a hybrid language implementation like Java and Python because of this performance difference.

That being said, most Ruby programs are not noticeably slower than their C++ counterparts. Instead of being worried about performance, "Matz" (the designer of Ruby), was more interested in making the programming activity less tedious. He

K.D. Lee, *Programming Languages*, DOI: 10.1007/978-0-387-79421-1_4,
© Springer Science+Business Media, LLC 2008

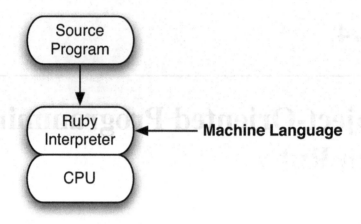

Fig. 4.1: Interpreting Ruby Programs

developed a language that allows programmers to program at a higher level and relieve themselves of many of the details of programming.

To write a Ruby program you write your code in a file or files of your choosing. In Linux you can run an interpreted program by naming the interpreter on the first line and making the file executable.

Example 4.1

In Linux and Unix there is a simple command called *echo* that will echo back to the screen the text you type. To write a similar tool, we can create a file called *recho* and enter this into the file.

```
#!/usr/local/bin/ruby
s = gets
puts s
```

The first line of the program may change slightly depending on where the Ruby interpreter is located. If you are not sure, you can type *which ruby* and Linux will tell you where the interpreter is located. After creating the file called *recho*, Linux needs to be told that the file is executable. To do this you enter the command

```
chmod +x recho
```

which tells Linux that echo is eXecutable. Then you can run this program by typing *recho* at the command line. That's it, you've written your first Ruby program. It doesn't work exactly like the Linux *echo* command, but it's close. You only have to make the file executable once. If you make more changes to the file, the program is still executable and you can change and try your program to your heart's content.

4.1 Designing Calc

As with the C++ version, we'll call the calculator interpreter **calc**. It will be invoked from the command-line as described in example 4.2. The interpreter will evaluate exactly the same expressions that the C++ version does.

Example 4.2

Here is a typical session with the completed calc interpreter.

```
%>calc
Please enter a calculator expression: (4S+5)*(7-R)
The result is 27
%>
```

The **S** represents the store operator. It stores the value that is computed to the left of it. In this case it stores a 4 in the memory location. The **R** represents the value that is stored in the memory location.

☞ Practice 4.1

Evaluate the following calculator expressions by hand.

1. (4+5)S*R
2. 3S + R
3. R + 3S
4. 2S*4 + R

While it is certainly possible to build the interpreter in an ad-hoc fashion, the goal of this project is to design the interpreter using a structured approach that will illustrate some of the concepts used in language implementation. In addition, some of the features of Ruby will be highlighted as this application is developed.

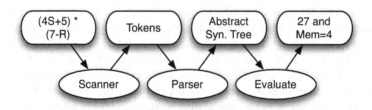

Fig. 4.2: Data flow through the Calc Interpreter

The calculator will read an expression from the command-line and process it as pictured in figure 4.2. The string read from the command-line is fed to a scanner which produces a tokenized representation of the string. The tokens are given to the parser which builds an abstract syntax tree (i.e. AST). The abstract syntax tree is then evaluated to produce the result.

The design of the code will closely follow the design in figure 4.2. Because we are using an object-oriented language, the first step is to identify the objects in the figure. There is a scanner, a parser, tokens, and an abstract syntax tree. The only thing left out is evaluate. Evaluate is something we do *to* the abstract syntax tree so we won't define that as an object.

There will also be one more object, a Calculator object that will start the evaluation of an expression. The calculator will have the memory location that we store numbers to and recall numbers from as part of its state.

Each of the objects identified above will be described by a class in Ruby. The following sections will cover the design and implementation for these classes. Along the way features of Ruby, and object-oriented languages in general, are described.

4.2   The Token Class

Each part of an expression that is read by the scanner becomes a token in the interpreter. The calculator language has several types of tokens. The complete list is *number*,+,-,*,/,(,),**S**,and **R**.

☞ Practice 4.2

> Identify the tokens in these expressions. Refer to the paragraph above to be sure you find them all.
>
> 1. 3S + R
> 2. (4+5)S*R

The token class will need to allow each of these tokens to be described. While in C++ an enumeration was used to describe the type of the tokens, Ruby provides a nice mechanism for defining symbols or atoms which are names that may be used within a program. A symbol is just a name preceded by a colon. So the names of the tokens are :number, :add, :sub, :times, :divide, :lparen, :rparen, :store, and :keyword. The :keyword token type will be used for **S** and **R**.

A token has a type. It should also have some extra information that will help should there be an error in the program or expression being evaluated. The line and column where the token began should also be stored with the token. The line will always be one in this program since the expression is one line long. GIven these requirements, we have enough information to specify what a Token object should look like.

Example 4.3

Here is the Ruby code representing token objects in the project. A Ruby class starts with the `class` keyword and ends with the keyword `end`. The constructor of a class is called `initialize`. Instance variables are preceded with @ when used in an expression or statement. The keyword `attr_reader` is something you likely haven't seen before. It is syntactic sugar. Ruby is designed as a language for programmers to relieve them of menial tasks. The `attr_reader` keyword will create accessor methods for the list of instance variables that follow it.

```ruby
class Token
    attr_reader :type, :line, :col

    def initialize(type,lineNum,colNum)
        @type = type
        @line = lineNum
        @col = colNum
    end
end
```

If you compare this code with the code for the C++ Token class given in examples 3.9 and 3.11 you will notice that there is already significantly less code in the Ruby version. There are several reasons for this. The lack of a garbage collector in C++ puts more burden on the programmer. A C++ programmer must write a destructor for a class when an object of that type points to a value that is heap allocated. Ruby is garbage collected, and therefore destructors are not relevant.

The Token class *declaration* and *implementation* must be written separately to allow for separate compilation. Since Ruby isn't compiled, it isn't necessary to write header files in Ruby to declare classes separately for use in other modules. In general, there will be significantly less code in the Ruby version of the interpreter.

The @ sign in the code above is used to distinguish a local variable from an instance variable. Instance variables in Ruby must be preceded by the @ to tell the Ruby interpreter that it is an instance variable. It is similar to the *this* pointer in C++ and Java and the *self* reference in the Python programming language.

4.3 Parameter Passing in Ruby vs C++

Because C++ allows parameters to be passed by value, pointer, or reference, it is sometimes necessary to declare functions to be **const** or constant, meaning that the function is an accessor method. Having more control over parameter passing requires the programmer do more of the work managing parameter passing. In Ruby there is only one way to pass parameters. They may only be passed by reference which is the way Java passes object parameters. Many interpreted, object-oriented languages implement only pass by reference since it is an efficient parameter passing mechanism. However, that also requires the programmer to remember that an object passed as a parameter may be mutated by the called function (if the object is mutable at all).

It should be noted that some of the objects in Ruby are immutable. For example, the integer class is immutable. When a class of objects are immutable it means that there are no methods to modify the data in the object once the object has been created. You can create new integer objects using other integer objects, but you can't change an integer object once it has been created.

When an object, or all objects of a class, are immutable, then the distinction of passing parameters by reference or by value is irrelevant. Either method has exactly the same outcome in terms of the output from the program. However, passing by reference is more efficient than passing by value because when passing by value the parameter must be copied. When the parameter is a large object, this may be quite a large amount of memory that has to be copied.

☞ Practice 4.3

Why doesn't the following code mutate the integer from 5 to 6?

```
x = 5
x = x + 1
```

4.4 Accessor and Mutator methods in Ruby

Given the attr_reader declaration in example 4.3 you can access the type instance variable of a token, t, by writing t.type. Accessor methods are all the same anyway, so letting Ruby provide the code for them is fine and certainly less tedious. Don't be tricked by this syntax. You cannot change an instance variable from outside the class using an attr_reader. If you intend to let code outside the class change an instance variable you must declare an attr_writer.

Each accessor function had to be written in the C++ version. The attr_reader and attr_writer syntactic sugar is not available in C++. Another difference involves the absence of the getLex method on the Token class. In the C++ version, the Token class included a getLex function even though there was no lexeme to return for general Tokens. In C++ it was necessary to write a getLex method for the Token class to avoid having to cast from a Token type to a LexicalToken type each time we wanted to call getLex. The reason we don't have to do this in Ruby will be explained in the next section.

In most languages an object is immutable if there exist no methods that mutate the object. For instance, the String class in Java is immutable. The String class in Ruby is not immutable. However, Ruby contains an interesting method defined on all objects called freeze that takes no parameters. When this method is called on an object it is frozen at run-time making any mutator method calls on the object throw an exception. By including a `freeze` method as part of the language implementation, objects can be made thread-safe. An immutable object can be safely accessed from more than one thread without the worry of a race condition between threads improperly updating the object.

4.5   Inheritance

In general the scanner will return tokens as described in the previous section. But, for a small number of tokens, numbers and keywords in particular, we need to know what number or keyword was found in the input. For these tokens we'll define a new LexicalToken class.

Example 4.4

Here is the code for the LexicalToken class. The `< Token` indicates that the class `LexicalToken` inherits from `Token`. Line 2 adds a `lex` accessor function. Line 5 calls the super class' constructor to initialize the inherited part of the object, while line 7 initializes the new part of the object.

```
1  class LexicalToken < Token
2     attr_reader :lex
3
4     def initialize(type,lex,lineNum,colNum)
5        super(type,lineNum,colNum)
6
7        @lex = lex
8     end
9  end
```

Inheritance allows us to define a new class in terms of an already defined class. In this case the LexicalToken class adds a lexeme to the Token class. *Inheritance* means that we don't have to write code again. In other words, inheritance helps with code reuse. Anywhere we write code that works with Tokens it will also work with LexicalTokens because a LexicalToken is a Token.

4.6   The AST Classes

According to the diagram in figure 4.2 the parser must build an abstract syntax tree of the expression to be evaluated. To do this a collection of classes must be declared. The AST classes can be designed as a hierarchy of classes where each type of AST class represents one type of node in an abstract syntax tree. If an evaluate method is defined for each type of node in an AST as well, then the evaluate methods can recursively compute the value associated with a calculator expression's AST.

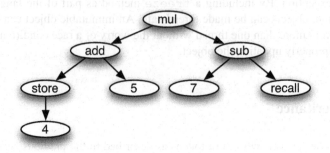

Fig. 4.3: Abstract Syntax Tree of expression in figure 3.1

Consider the abstract syntax tree in figure 4.3. There are mul, add, sub, store, recall, and number nodes within the tree. If we abstract away from the details a bit we also see there are some nodes which have no children, some with one child, and some with two children.

The tree in figure 4.3 can be evaluated to produce the value of the expression. By recursively calling the evaluate method associated with each node in the AST, the tree is traversed. A postfix traversal of the tree yields the value where the postfix operation is the calculation associated with the AST node being visited. For instance consider the steps in the postfix traversal given here.

1. The traversal begins by recursively descending the left side of the tree down to the 4 node. Visiting that node returns 4.
2. The store node takes the 4 and stores it in the calculator's memory. It also returns the 4.
3. The add can't be visited yet since it has a right child (the 5). The 5 node is visited and returns the 5.
4. The add node can now be visited. It take the 4 and the 5, adds them together and returns the 9.
5. The mul node can't be visited until its right child is visited. Postorder traversal of the sub node calls the traversal on the 7 node, which returns 7.
6. The sub node still can't be visited yet. The recall node is traversed and returns the value in the calculator memory, the 4.
7. The 7-4 is computed by visiting the sub node and returns 3.
8. Visiting the mul node computes 9*3 or 27.

The evaluation procedure can be accomplished by writing a polymorphic evaluate method for an abstract syntax tree. In the C++ version of this code we first had to define an abstract base class called AST that defined the methods that would be common to all the subclasses. This isn't necessary in Ruby. Using Ruby it is never necessary to create an abstract base class because Ruby is dynamically typed instead of being statically typed like C++. The next section describes the difference between dynamic and static typing when it comes to polymorphism.

The AST classes still use inheritance to take advantage of code reuse. Some nodes in the tree are unary nodes while others are binary nodes. To avoid rewriting the code to manage the subtrees, a UnaryNode and BinaryNode class are defined along with the accessor methods for accessing their subtrees and constructors for constructing unary and binary nodes.

Example 4.5

Here is the code for the two base classes, UnaryNode and BinaryNode.

```
class BinaryNode
   attr_reader :left, :right

   def initialize(left,right)
      @left = left
      @right = right
   end
end

class UnaryNode
   attr_reader :subTree

   def initialize(subTree)
      @subTree = subTree
   end
end
```

The code from the UnaryNode and BinaryNode classes can be reused in defining AST nodes involving unary and binary operations. For instance, the addition, subtraction, multiplication, and division operations are all binary operations and hence their AST classes inherit from the BinaryNode class. The store operation is a unary operation so its class should inherit from the UnaryNode class.

Example 4.6

Here is the code for the AddNode and SubNode classes. The other classes are left as an exercise.

```
class AddNode < BinaryNode
   def initialize(left, right)
      super(left,right)
   end

   def evaluate()
      return @left.evaluate() + @right.evaluate()
   end
end

class SubNode < BinaryNode
   def initialize(left, right)
      super(left,right)
   end
```

```
15
16    def evaluate()
17        return @left.evaluate() - @right.evaluate()
18    end
19  end
```

☞ Practice 4.4

Write the NumNode class for AST nodes that contain numbers.

After looking at the code in examples 4.5 and 4.6 you may have noticed that there is no common ancestor class containing an `evaluate` method. The evaluate method had to be defined in a common ancestor class in C++ for polymorphism to work. This is not true in Ruby. Polymorphism works differently in Ruby. This is the topic of the next section.

4.7 Polymorphism in Ruby

Ruby is a dynamically typed language. This means that types are not determined at compile-time, since Ruby is not compiled. In fact, the type of an expression in Ruby is not determined until the program is run. This has some consequences for the way we write code in Ruby. There are two problems with run-time type checking. Checking types at run-time slows down the execution of the program, although not significantly for most applications. The second problem relates to testing code. If a program has a type error, run-time type checking doesn't detect it until the program evaluates the offending expression in the code. Since some errors may be on obscure paths through the code, you have to test your code very thoroughly to be sure you have found all the type errors in your program.

There are advantages to dynamic type checking. The biggest advantage may be in the way polymorphism works. Polymorphism is the ability to pick the right method to invoke depending on the object the method is being invoked upon. The project presented in this chapter makes use of polymorphism when an abstract syntax tree (AST) is evaluated.

Consider the abstract syntax tree in figure 4.3 on page 98. To evaluate the tree, the `evaluate` method is called on the root's `MulNode` object. The MulNode evaluate method polymorphically calls the evaluate method on the left and right subtrees like the AddNode's evaluate method shown on page 99. How does the correct evaluate get called? There isn't anything in the MulNode class that says the call to the left subtree's evaluate method should be a call to the AddNode's evaluate. Of course, many times the left subtree of a MulNode won't be an AddNode. The information about which evaluate method is called must be in the object.

Figure 4.4 depicts the organization of objects and classes within Ruby. Each object in Ruby contains a pointer to its corresponding class (the bold dashed lines in

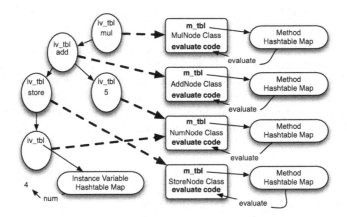

Fig. 4.4: Ruby object organization

the figure). Within each object is a field called iv_tbl. This field points to a hash table, one per object, that points to the instance variables of the object. A hash table is a data structure that can map one object to another object. Each iv_tbl maps strings representing the names of the instance variables (hence the iv_tbl name) to the values stored in those instance variables. Only one instance variable hash table was shown in figure 4.4 because the picture would be too cluttered otherwise, but there is one instance variable hash table per object. On average, a hash table can be used to look up a mapped value in constant time so having to look up instance variables each time they are accessed isn't too bad.

Each object also has a pointer to its class. Likewise, each class has a hash table mapping method names to the code that implements them. Each time evaluate is called on an object, the m_tbl hash table maps the name "evaluate" to the code that implements it. This is all done at run-time. Polymorphism works in Ruby by dynamically looking up the right method in the hash table. If the method is not found in the hash table then there is a type error.

Since hash tables have an amortized lookup complexity of constant time, all the lookups of methods and instance variables in Ruby is not a huge penalty. Looking up each instance variable and method when it is used means that objects are completely dynamically typed in Ruby.

☞ Practice 4.5

Recalling that every value in Ruby is an object, draw a picture of the store node object shown in figure 4.4 given what you now know about objects in the Ruby model. In your picture draw all the objects that the store node refers to in this example. Be careful when you do this. Remember that every value is an object in Ruby.

4.8 The Scanner

Referring back to figure 4.2 the scanner reads characters from the input and builds
Token objects that are used by the parser. To accomplish this the scanner needs
to read characters from a stream and decide how to group them into tokens. The
parser will get tokens from the scanner by calling a getToken method. Sometimes
the parser gets a token and needs to put it back to get again later. In that case a
putBackToken method will put back the last token that was returned by getToken.

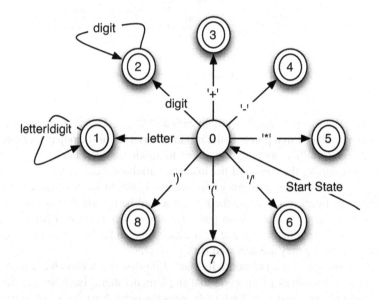

Fig. 4.5: A Finite State Machine for the Scanner

Internally, the scanner object is a finite state machine. A finite state machine
(fsm) consists of a set of states and a set of transitions from one state to another
based on the current character in the input. Figure 4.5 defines the Scanner's finite
state machine. The fsm starts in state zero, reads one character and transitions to
one of the eight states depending on the character. In state one there is a transition
that stays in state one as long as a letter or digit is read. The fsm continues to read
characters and make transitions until a character appears that has no transition from
the current state. At that point, if the state is an accepting state (i.e. a double circle
state) the string of characters is recognized as a token.

This fsm reads a token and returns it. The EOF and keyword tokens are handled
as special cases. EOF is handled in state zero when the end of stream is reached.
State one compares the identifier to a list of keywords to see if it should return

a keyword or identifier token. The complete implementation can be found in appendix B.

Whitespace is read and thrown away by the scanner. Whitespace consists of blanks, tabs, and newline characters. When a character is read that is not recognized by a transition from the current state, the fsm returns the current token (since states 1-8 are all accepting states) saving the current character for later. The next call to getToken resumes with the current character. If there is no valid transition from state zero on the current character the fsm returns the character as an unrecognized token.

An fsm is a model of computation for recognizing strings of characters. An fsm is easily implemented using a while loop, a case statement, and one variable to record the current state. Fsm's are used in many contexts including network protocol implementations, pattern recognition, simulations, and of course language implementation. There are tools to build powerful fsm's. However, it's good to see a hand-written one to aid in understanding some of the theory behind fsm's, too.

4.9 The Parser

Figure 4.2 shows the parser reading tokens and producing an abstract syntax tree as its output. In section 2.7 on page 31 a parser was defined as a program that given a sentence (i.e. a string of tokens), checks to see if the sentence is in the language of a given grammar. The parser that is discussed in this section is a top-down parser. The parser will build the AST from the top-down. In reality, top-down is a bit of a misnomer. While the construction of the tree starts at the top, the tree is actually built bottom-up by a recursive descent of the tree. That's why top-down parsers are also called recursive descent parsers.

To begin to design a parser there must be a grammar to model it after.

Example 4.7

This is the Calculator language's grammar.

$Prog \rightarrow Expr\ EOF$
$Expr \rightarrow Expr + Term \mid Expr - Term \mid Term$
$Term \rightarrow Term * Storable \mid Term/Storable \mid Storable$
$Storable \rightarrow Factor\ S \mid Factor$
$Factor \rightarrow number \mid R \mid (Expr)$

A recursive descent parser is recursive as the name suggests. The implementation of the parser is given to us by its grammar. In the implementation, each nonterminal becomes a function in the parser. Each rule in the grammar is part of a function that is named by the nonterminal on the left side of the arrow in the rule. In the grammar above each line would correspond to a function in the parser. Each appearance of a nonterminal on the right hand side of a production is a function call. Each appearance of a token on the right hand side of a production is a call to the scanner to get a token. From this definition, writing the parser is pretty straightforward.

A First Attempt at Writing the Parser

The parser will read the tokens and build an abstract syntax tree like the one figure 4.3. To write the top-down parser of these expressions each nonterminal becomes a function. The grammar dictates how to write the parser. The body of each function is given by the right hand side of its corresponding production.

Example 4.8

The Prog and Expr functions for the Parser.

```
1    def Prog()
2        result = Expr()
3        t = @scan.getToken()
4
5        if t.type != :eof then
6            print "Expected EOF. Found ", t.type, ".\n"
7            raise "Parse Error"
8        end
9
10       return result
11   end
12
13   def Expr()
14       ast = Expr()
15       t = @scan.getToken()
16       ...
17   end
```

There is a big problem with the Expr function given above. It is recursive and there is no base case. This means if you call the Expr function, it will go into infinite recursion resulting in run-time stack overflow. The grammar given above isn't suited for top-down parsing.

A Better Attempt at Writing a Top-Down Parser

The problem in the previous section is that the grammar is not LL(1). For a grammar to be LL(1) means that the choice of which production to apply next in a left-most derivation of a sentence can be made by looking ahead at the next token. The number one in LL(1) means that only one token of lookahead is needed to decide which production to use. Although the grammar above is LALR(1), it is not appropriate for constructing a recursive descent parser. An LL(1) grammar is needed to build a recursive descent or top-down parser. An LALR(1) grammar is a grammar that can be given to a program to construct a reverse right-most derivation of a sentence in the grammar looking ahead at only the next token in the sentence. This is what a bottom-up parser does and bottom-up parser generators can take a grammar like the one above and automatically construct a parser for it. Because bottom-up parsers are harder to write, we usually rely on a parser generator program to write the parser for us when generating a bottom-up parser.

Top-down parsers are much simpler to write and are typically written by hand. However, to create a top-down parser you have to have an LL(1) grammar. Fortunately, it is relatively easy to convert an LALR(1) grammar to an LL(1) grammar. There are two steps involved.

1. Eliminate left recursion.
2. Perform left factorization where appropriate.

Eliminate Left Recursion

Eliminating left recursion means eliminating rules like $Expr \rightarrow Expr + Term$. Rules like this are left recursive because the *Expr* function would first call the *Expr* function in a recursive descent parser as in example 4.8 above. Without a base case first, we are stuck in infinite recursion (a bad thing). To eliminate left recursion we look to see what Expr can be rewritten as when deriving a sentence. In this case, Expr can only be replaced by a Term so we replace Expr with Term in the productions. Then, we add a new nonterminal to represent the rest of production from the LALR(1) grammar. In this case, the + *Term* and the - *Term* are left after we replace the initial *Expr* in the productions in the grammar above. The usual way to eliminate left recursion is to introduce a new nonterminal to handle all but the left recursive nonterminal. Two rules in the grammar are left recursive and must be rewritten.

Example 4.9

An LL(1) Calculator Language Grammar

$Prog \rightarrow Expr\,EOF$
$Expr \rightarrow Term\,RestExpr$
$RestExpr \rightarrow +\,Term\,RestExpr \mid -\,Term\,RestExpr \mid <null>$
$Term \rightarrow Storable\,\,RestTerm$
$RestTerm \rightarrow *\,Storable\,RestTerm \mid /\,Storable\,RestTerm \mid <null>$
$Storable \rightarrow Factor\,S \mid Factor$
$Factor \rightarrow number \mid R \mid (Expr)$

In this example the $Expr \rightarrow Expr + Term \mid Expr - Term \mid Term$ is replaced by the second and third lines of the grammar given above. Likewise, the left recursion in $Term \rightarrow Term * Storable \mid Term/Storable \mid Storable$ is rewritten as the fourth and fifth lines of the grammar above.

Perform Left Factorization

Left factorization isn't needed on this grammar so this step is skipped. Left factorization is needed when the first part of two or more productions is the same and the rest of the similar productions are different. Left factorization is important in languages like Prolog because without it the parser may have to backtrack. Since backtracking won't work when reading something from an input stream you must perform left factorization by writing a new rule that handles the common prefix of the two offending rules. However, it isn't needed in Ruby if you recognize the common prefix and code the function appropriately.

Translating the LL(1) Grammar to Ruby

Once you have an LL(1) grammar you use it to build a parser as follows. The following construction causes the parser to return an abstract syntax tree for the sentence being parsed.

1. Construct a function for each nonterminal. Each of these functions should return a node in the abstract syntax tree.
2. Depending on your grammar, some nonterminal functions may require an input parameter of an abstract syntax tree (ast) to be able to complete a partial ast that is recognized by the nonterminal function.
3. Each nonterminal function should call getToken on the scanner to get the next token as needed. If after getting the token the code determines it didn't need the token after all, the nonterminal function should call the scanner's putBackToken function to put back the token. If the parser is based on an LL(1) grammar, it should never have to put back more than one token at a time.
4. The body of each nonterminal function is a series of if statements that choose which production to expand upon depending on the value of the next token. The body of the function is determined by the productions of the grammar with the given nonterminal on the left hand side of the arrow.

The construction above is very simple, but can be confusing without an example. Consider the LL(1) grammar given above. Assume that you have two classes called AddNode and SubNode that are derived from the BinaryNode class.

Example 4.10

The Parser's Prog and Expr Functions

```
1    def Prog()
2        result = Expr()
3        t = @scan.getToken()
4
5        if t.type != :eof then
6            print "Expected EOF. Found ", t.type, ".\n"
```

```
 7            raise "Parse Error"
 8         end
 9
10         return result
11     end
12
13     def Expr()
14         return RestExpr(Term())
15     end
```

The code in example 4.10 raises an exception if an error is discovered during parsing. You would normally take appropriate action during error conditions, but raising an exception is a legitimate way to deal with a parsing problem. The Prog function returns a reference to an AST, which is the abstract syntax tree representing the expression that was parsed.

The Expr function corresponds to the *Expr* rules in the grammar in example 4.9. The rule says to first call the Term function. The result of calling this function is an AST (as all nonterminal functions return an AST).

☞ Practice 4.6

The RestExpr function is slightly different from the Prog and Expr functions. The RestExpr function has an AST parameter which we'll call e. The RestExpr function first gets a token and then decides what to do based on that token. If it is an **add** token it builds a new AST AddNode with the part of the tree given to it (i.e. e) as the left subtree and the result of calling Term as the right subtree. The subtract AST nodes are handled similarly. Otherwise, there wasn't a token that the RestExpr knows about so the token is put back and the AST e is returned as its AST.

Write the RestExpr function described here. Remember you can refer to the grammar in example 4.9.

The remainder of the parser implementation can be patterned after the code in example 4.10 using the grammar in example 4.9 as a guide. The remainder of the code is left as an exercise.

4.10 Putting It All Together

One more class is required to tie together the pieces that have been developed in this chapter. The Calculator class contains a memory location that can hold a stored value. The value can also be retrieved on demand. The calculator can evaluate an expression that is given to it as a string.

There is only one calculator object in this program and it will be useful if it is declared as a global variable. The $calc global variable is needed in the abstract syntax tree implementation so the AST can have access to the calculator's memory location. By declaring $calc with a dollar sign we tell Ruby that calc is a global variable that can be accessed from any class and method. Global variables are generally a bad idea. This is one case where it is justified. Without it just about every object presented in this chapter would have to keep a reference to the calculator. That's a lot of overhead to have access to one little memory location.

Example 4.11

The Calculator class implementation

```
1  class Calculator
2      attr_reader :memory
3      attr_writer :memory
4
5      def initialize()
6          @memory = 0
7      end
8
9      def eval(expr)
10         parser = Parser.new(StringIO.new(expr))
11         ast = parser.parse()
12         return ast.evaluate()
13     end
14  end
```

The Calculator evaluates an expression by creating a StringIO object over the string containing the expression. A StringIO is an input stream constructed from a string. This stream is passed to the Parser constructor. The Parser in turn constructs a Scanner object over the stream to get the tokens from the string.

If all goes well, the string is parsed and the parser returns an AST. The tree is then evaluated (polymorphically) to yield the result. The result is returned to the main program to be printed. This flow of data is depicted in the dataflow diagram shown in figure 4.2.

Example 4.12

The code to start it all.

```
1  text = gets
2  $calc = Calculator.new()
3
4  puts "The result is " + $calc.eval(text).to_s
```

The global variable is declared in the top-level code. This code gets the input line from the user, creates the calculator, and calls eval on the calculator giving it the input line. The result is printed to the screen.

4.11 Static vs Dynamic Type Checking

In this chapter we have learned how to design and implement a calculator in Ruby. In the last chapter the same project was tackled in C++. While the two projects have many similarities, there are important differences between them as well. The primary difference between the two projects stems from the way type checking is handled in the two languages and how polymorphism is implemented.

These differences can be highlighted if we examine the AST code for the two calculator implementations. In C++, the AST classes must all inherit from a common ancestor, the *AST* class.

Example 4.13

Here is the AST, BinaryNode, UnaryNode, and AddNode class declarations in C++.

```
 1   class AST {
 2   public:
 3       AST();
 4       virtual ~AST() = 0;
 5       virtual int evaluate() = 0;
 6   };
 7
 8   class BinaryNode : public AST {
 9   public:
10       BinaryNode(AST* left, AST* right);
11       ~BinaryNode();
12
13       AST* getLeftSubTree() const;
14       AST* getRightSubTree() const;
15   private:
16       AST* leftTree;
17       AST* rightTree;
18   };
19
20   class UnaryNode : public AST {
21   public:
22       UnaryNode(AST* sub);
23       ~UnaryNode();
24
25       AST* getSubTree() const;
26   private:
27       AST* subTree;
28   };
29
30   class AddNode : public BinaryNode {
31   public:
32       AddNode(AST* left, AST* right);
33
34       int evaluate();
35   };
```

The inheritance is required so the right evaluate method will be called polymorphically. When the AST node is actually an *AddNode* the *AddNode* evaluate will be called. When the node is a *SubNode* the *SubNode* evaluate will be called because of polymorphism. This is what we would like when a AST is evaluated. The right *evaluate* methods get called depending on the types of nodes that make up the tree.

For this code to compile in C++ all nodes must inherit from a common ancestor. This is because the parser returns an AST node and C++ checks the types of each possible AST node to be sure it inherits from *AST* either directly or indirectly. C++ requires this because for polymorphism to work the compiler needs to be able to locate the evaluate method in the **vtable** for each possible node in a tree. By requiring a common ancestor, the compiler can guarantee that the *evaluate* method will always be at the same location in the **vtable**.

Contrast this to the way Ruby works. In the Ruby AST implementation there is no common ancestor. Inheritance is used a little, but only for code reuse. Inheritance is not needed for polymorphism.

Example 4.14

Here is the equivalent code in Ruby. Notice there is no common ancestor of the AST classes.

```ruby
class BinaryNode
    attr_reader :left, :right

    def initialize(left,right)
        @left = left
        @right = right
    end
end

class UnaryNode
    attr_reader :subTree

    def initialize(subTree)
        @subTree = subTree
    end
end

class AddNode < BinaryNode
    def initialize(left, right)
        super(left,right)
    end

    def evaluate()
        return @left.evaluate() + @right.evaluate()
    end
end
```

Ruby does not check the types of expressions in the program before executing the code. This isn't necessary in Ruby because all methods are looked up via a hash table at run-time as described in this chapter. Since no type checking is done prior to executing the code, when evaluate is called to compute the value represented by an AST, the right *evaluate* methods are looked up at run-time and the correct code gets called. When evaluate is called on an *AddNode* the *AddNode* evaluate code is looked up and executed.

The difference in how polymorphism is implemented in C++ and Ruby has some pretty big consequences. The Ruby code is substantially shorter than the C++ code and it is certainly more convenient to write the Ruby program since all the extra classes and syntax aren't required. However, errors in type won't show up in a Ruby program until the program executes the code with the error in it. In C++, if we forget to implement an evaluate method for one of the AST classes we'll find out when we compile the program. In Ruby we wouldn't find out until we actually tried to evaluate a tree containing one of those nodes.

Static typing insures that most type errors are found when the program is compiled. Static typing requires more work to maintain the type hierarchies. Dynamic typing requires less coding but means that errors may not be found until run-time.

4.12 Exercises

1. What's the value of (R+7)/4S if the memory contained 4 prior to evaluating this expression?
2. What is the value of the memory location after evaluating the previous expression?
3. What does the abstract syntax tree look like for the expression (R+7)/4S?
4. How could the calculator language be modified to allow more than one memory location like modern calculators? Discuss what changes would be required to implement this enhanced calculator language.
5. Complete the calculator interpreter by downloading the code given in this chapter and finishing the implementation of the parser and the AST in the calc file. The rest of the project is provided.

When you download the code you will want to unzip the package with some sort of unzip program. On Linux you can issue the command,

```
unzip rubcalc.zip
```

Then you can make the program and run it. Here is an example of making and running you can use to get started.

```
$ unzip cppcalc.zip
$ cd rubycalc
$ calc
Please enter a calculator expression: 5 + 4
The result is 9
$
```

Commands that you enter are preceded by a dollar sign. The program will add two integers together as provided. Your job is to extend the project to the full calculator language. This requires changes to the parser and ast modules. The parser changes are highlighted in section 4.9. You can complete the parser by completing the functions that are incomplete in the parser class. These functions can be patterned after the code presented in the chapter.

The parser code will require that you build AST nodes for storing values and for recalling values from the calculator's memory. You will also need multiply and divide nodes in the abstract syntax tree. These new node types can be added to the AST code using the existing code as a pattern.

The store and recall nodes in the AST will need to access the memory location of the calculator. The global variable called **$calc** can be used to access the calculator's memory. The line

```
$calc.memory = 6
```

will store 6 in the calculator's memory. Similarly, the expression $calc.memory will retrieve the value stored in the memory of the calculator.

6. Once you have completed the project described above extend the calculator language to allow more than one memory location to hold a value.

7. Modify the project to be a compiler instead of an interpreter. Instead of evaluating the expression, generate EWE code for it instead.

   In addition, to make this interesting, add a new keyword to the language, called *I*, that when executed waits for user input before proceeding. The value returned by the call to *I* is the value entered at the keyboard.

   This project can be implemented with a few modifications. First, the eval function of the abstract syntax tree will print code to a file called *a.ewe* instead of directly evaluating the expression. The web page for the text contains a link to code to start this modified project. Remember, you are now printing code in this project and not evaluating. The EWE interpreter is evaluating the code.

   For this project to work well you should decide on a model of computation for the generated EWE code to follow. A stack makes a nice model. When you generate EWE code for an expression, the resulting value should always be left on the top of a stack that you simulate using the EWE interpreter. That way, you can always find a value when you need it. Consult the code provided on the web site to see how this stack is simulated. The code provided has enough of the compiler implemented to add two integers together.

4.13 Solutions to Practice Problems

These are solutions to the practice problems . You should only consult these answers after you have tried each of them for yourself first. Practice problems are meant to help reinforce the material you have just read so make use of them.

Solution to Practice Problem 4.1

1. 81
2. 6
3. Depends on the initial value of memory. It could be an error. Assuming the calculator starts with 0 in memory the answer would be 0. If the calculator is written to evaluate more than one expression in a session then the memory might contain the last value stored.
4. 10

Solution to Practice Problem 4.2

1. There are number, keyword, add, keyword tokens in this one.
2. The tokens are: lparen, number, add, number, rparen, keyword, times, keyword.

Solution to Practice Problem 4.3

The integer object that x refers to (i.e. the 5) is not mutated. The reference x is changed to point to a new object, the result of adding 5 and 1.

Solution to Practice Problem 4.4

This is the NumNode implementation. Notice it does not inherit from anything because it is a leaf node and not a UnaryNode or BinaryNode. Inheritance is only used for code reuse in Ruby. It is not needed for polymorphic type checking.

```ruby
class NumNode
    def initialize(num)
        @num = num
    end

    def evaluate()
        return @num
    end
end
```

Solution to Practice Problem 4.5

This is how Ruby objects are organized in memory. This is still only a sampling of the memory organization. There are details omitted because there is too much to display the complete organization of even these two objects. It should give you a good idea of the organization with Ruby though.

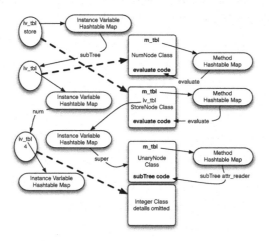

Fig. 4.6: Ruby object organization

Solution to Practice Problem 4.6

This is the RestExpr implementation.

```
def RestExpr(e)
    t = @scan.getToken()

    if t.type == :add then
        return RestExpr(AddNode.new(e,Term()))
    end

    if t.type == :sub then
        return RestExpr(SubNode.new(e,Term()))
    end

    @scan.putBackToken()

    return e
end
```

4.14 Additional Reading

There is a lot of excellent material on Ruby on the web. The web site www.ruby-doc.org is a good place to start. Programming Ruby[35] is a good reference for both learning and looking up information if you prefer a book to the web. There are many other good reference books that are available as well. Ruby on Rails is a useful tool for quickly prototyping database code. When properly configured, rails will provide you with a web page front-end to a database application. Generated code is easy to understand and easy to modify to suit your needs.

Chapter 5

Functional Programming in Standard ML

As you might guess by the title, functional programming has something to do with programming with functions. However, what the title, **Functional Programming**, doesn't say is what functional programming languages lack. Specifically, pure functional languages lack assignment statements and iteration. Iteration relates to the ability to iterate or repeat code as in a loop of some sort. It is impossible in a pure functional language to declare a variable that gets updated as your program executes! If you think about it, if there are no variables, then there isn't any reason for a looping construct in the language. Iteration and variables go hand in hand. But, how do you get any work done without variables? The primary mode of programming in a functional language is through recursion.

Functional languages also contain a feature that other languages don't. They allow functions to be passed to functions as parameters. We say that these functions are higher-order. Higher-order functions take other functions as parameters and use them. There are many useful higher order functions that are derived from common patterns of computation. Particular instances of these patterns commonly have one small difference between them. If that small difference is left as a function to be defined later, we have one function that requires another function to complete its implementation. Higher-order functions may be customized by providing some of their functionality later. In some ways this is the functional equivalent of what inheritance or interfaces provide us in object-oriented languages.

These two features, lack of variables and higher-order functions, drastically change the way in which you think about programming. Programming recursively takes some time to get used to, but in the end it is a very nice way to program. Programming recursively is more declarative. Writing imperative programs is prescriptive. When programming declaratively we can focus on *what* we want to say about a problem instead of exactly *how* to solve a problem.

But why would we want to get rid of variables in a programming language? The problem is that variables often make it hard to reason about our programs. Functional languages are more mathematical in nature and have certain rules like

K.D. Lee, *Programming Languages*, DOI: 10.1007/978-0-387-79421-1_5,
© Springer Science+Business Media, LLC 2008

commutativity and associativity that they follow. Rules like associativity and commutativity can make it easier to reason about our programs.

☞ Practice 5.1

Is addition commutative in C++, Pascal, or Java? Will `write(a+b)` always produce the same value as `write(b+a)`? Consider the follow Pascal program (or almost a Pascal program, anyway).

```
1   program P;
2      var b : integer;
3
4      function a() : integer;
5      begin
6         b:=b+2;
7         return 5
8      end;
9   begin
10     b:=10;
11     write(a()+b)          (* or write(b+a()) *)
12  end.
```

What does this program produce? What would it produce if the statement were `write(b+a())`?

5.1 Imperative vs Functional Programming

You are probably familiar with at least one imperative language. Languages like C, C++, Java, Python, and Ruby are considered imperative languages because the fundamental construct is the assignment statement. In each of these languages we declare variables and assign them values, updating those variables as a program's execution progresses.

Imperative languages are heavily influenced by the von Neumann architecture of computers that includes a store and an instruction counter; the computation model has control structures that iterate over instructions that make incremental modifications of memory. Assignment of values to variables, for loops, and while loops are all part of imperative languages. The principal operation is the assignment of values to variables. Programs are statement oriented, and they carry out algorithms with statement level sequential control. In other words, computing is done by side-effects.

Sometimes problems with imperative programs stem from these side-effects. It is difficult to reason about a program that relies on side-effects. If we wish to reuse the code of an imperative program then we must be sure that the same conditions are true before the reused code executes since imperative code relies on a certain machine state. As programmers we sometimes forget which preconditions are required and what the postconditions of executing a segment of code might be and that can lead to bugs in our programs.

Functional languages are based on the mathematical concept of a function and do not reflect the underlying von Neumann architecture. These languages are concerned with data objects and values instead of variables. The principal operation is function application. Functions are treated as first-class objects that may be stored in data structures, passed as parameters, and returned as function results. Primitive functions are generally supplied with the language implementation. Functional languages allow new functions to be defined by the programmer.

Functional program execution consists of the evaluation of an expression, and sequential control is replaced by recursion. There is no assignment statement. Values are communicated primarily through the use of parameters and return values. Without variables, loop statements don't have a purpose and so they also don't exist in pure functional languages.

Pure functional languages have no side-effects. If input and output are considered side-effects then the only pure functional programs are those that read no input and produce no output. In other words, according to this definition, the only pure functional programs are those that do nothing! Realistically, side-effects are avoided.

Scheme is generally considered a pure functional language even though it does include input and output as part of its definition. In general, pure functional languages like Scheme avoid or at least isolate code that performs input and output operations. More importantly, input and output operations in functional languages do not update the state of variables within a program. The only state update relates to the state of the stream of characters being read from or written to.

What is amazing is that it has been proven that exactly the same things can be computed with functional languages as can be computed with imperative languages. This is known because a Turing machine, the theoretical basis for imperative programming and the design of the computer, have been proven equivalent in power to the Lambda Calculus, the basis for all functional programming languages.

5.2 The Lambda Calculus

All functional programming languages are derived either directly or indirectly from the work of Alonzo Church and Stephen Kleene. The lambda calculus was defined by Church and Kleene in the 1930's, before computers existed. At the time, mathematicians were interested in formally expressing computation in some written form other than English or other informal language. The lambda calculus was designed as a way of expressing those things that can be computed. It is a very small, functional programming language. In the lambda calculus, a function is a mapping from the elements of a domain to the elements of a codomain given by a rule. Consider the function $cube(x) = x^3$. What is the value of the identifier $cube$ in the definition $cube(x) = x^3$? Can this function be defined without giving it a name?

$\lambda x.x^3$ defines the function that maps each x in the domain to x^3. We can say that this definition or *lambda abstraction*, $\lambda x.x^3$, is the value bound to the identifier *cube*. We say that x^3 is the *body* of the lambda abstraction. Every lambda *abstraction* in lambda notation is a function of one identifier. However, lambda *expressions* may contain more than one identifier.

Example 5.1

The expression $y^2 + x$ can be expressed as a lambda abstraction in one of two ways:

$$\lambda x.\lambda y.y^2 + x$$
$$\lambda y.\lambda x.y^2 + x$$

In the first lambda abstraction the x is the first parameter to be supplied to the expression. In the second lambda abstraction the y is the parameter to get a value first. In either case, the abstraction is often abbreviated by throwing out the extra λ. In abbreviated form the two abstractions would become $\lambda xy.y^2 + x$ and $\lambda yx.y^2 + x$.

Normal Form

To say the lambda calculus, or any language, has a normal form means that each expression that can be reduced has a simplest form. It means that we can reduce more complex expressions to simpler expressions in some mechanical way. The lambda calculus exhibits a property called *confluence*.

Confluence means that one or more reduction strategies (or intermixing them) always leads to the same normal form of an expression, assuming the expression can be reduced by the reduction strategy. This property of confluence was proven in the Church-Rosser theorem.

Function application (i.e. calling a function) in lambda notation is written with a lambda abstraction followed by the value to call the abstraction with. Such a combination is called a *redex*.

Example 5.2

To call $\lambda x.x^3$ with the value 2 for x we would write

$$(\lambda x.x^3)2$$

This combination of lambda abstraction and value is called a *redex*.

A redex is a lambda expression that may be reduced. Typically a lambda expression contains several redexes that may be chosen to be reduced. Function application is left-associative meaning that if more than one redex is available at the same level of parenthetical nesting, the left-most redex must be reduced first. If the left-most outer-most redex is always chosen for reduction first, the order of reduction is called normal order reduction. When a redex is reduced by applying the lambda calculus equivalent of function application it is called a β-reduction (pronounced beta-reduction).

Example 5.3

This is the normal order reduction of $(\lambda xyz.xz(yz))(\lambda x.x)(\lambda xy.x)$. The redex to be β-reduced at each step is underlined.

$(\lambda xyz.xz(yz))(\lambda x.x)(\lambda xy.x)$
$\Rightarrow \underline{(\lambda yz.(\lambda x.x)z(yz))(\lambda xy.x)}$
$\Rightarrow \lambda z.\underline{(\lambda x.x)z}((\lambda xy.x)z)$
$\Rightarrow \lambda z.z(\underline{(\lambda xy.x)z})$
$\Rightarrow \lambda z.z(\lambda y.z)\square$

☞ Practice 5.2

Another reduction strategy is called applicative order reduction. Using this strategy, the left-most inner-most redex is always reduced first. Use this strategy to reduce the expression in example 5.3. Be sure to parenthesize your expression first so you are sure that you left-associate redexes.

In practice problem 5.2 you should have reduced the lambda expression to the same reduced lambda expression seen in example 5.3. If you didn't, you did something wrong. If you want more experience with reducing lambda expressions you may wish to consult a lambda expression interpreter. One excellent interpreter was written by Peter Sestoft and is available on the web. It is located at http://www.dina.dk/ sestoft/lamreduce/.

Problems with Applicative Order Reduction

Sometimes, applicative order reduction can lead to problems. For instance, consider the expression $(\lambda x.y)((\lambda x.xx)(\lambda x.xx))$.

☞ Practice 5.3

Reduce the expression $(\lambda x.y)((\lambda x.xx)(\lambda x.xx))$ with both normal order and applicative order reduction. Don't spend too much time on this!

This practice problem shows why the definition of confluence above includes the phrase, *assuming the expression can be reduced by the reduction strategy*. Applicative order may not always result in the expression being reduced. No fear, if that happens we are free to use normal order reduction for a while since intermixing reduction strategies will not affect whether we arrive at the normal form for the expression or not.

5.3 Getting Started with Standard ML

Standard ML (or just SML) is a functional language based on Lisp which in turn is based on the lambda calculus. Important ML features are listed below.

- SML is higher-order supporting functions as first-class values.
- It is strongly typed like Pascal, but more powerful since it supports polymorphic type checking. With this strong type checking it is pretty infrequent that you need to debug your code!! What a great thing!!!
- Exception handling is built into Standard ML. It provides a safe environment for code development and execution. This means there are no traditional pointers in ML. Pointers are handled like references in Java.
- Since there are no traditional pointers, garbage collection is implemented in the ML system.
- Pattern-matching is provided for conveniently writing recursive functions.
- There are built-in advanced data structures like lists and recursive data structures.
- A library of commonly used functions and data structures is available called the **Basis Library**.

There are several implementations of Standard ML. Standard ML of New Jersey and Moscow ML are the most complete and certainly the most popular. There is also a SML.NET implementation that targets the Microsoft .NET run-time library and can be integrated with other .NET languages. There is an MLj implementation that targets the Java Virtual Machine. Poly/ML is another implementation that includes support for Windows programming. While many implementations exist, they all support the same definition of SML. If you write a Standard ML program that runs in one environment, it'll run on any other implementation as long as you are not using platform specific functions.

SML has been successfully used on a variety of large programming projects. It was used to implement the entire TCP protocol [3] on the FOX Project. It has been used to implement server side scripting on web servers. It was originally designed as a language to write theorem provers and has been used extensively in this area. It has been used in hardware design and verification. It has also been used in programming languages research.

The rest of this chapter will introduce you to SML. By the end of the chapter you should understand and be able to use many of the important features of the language. This text is based on the Standard ML of New Jersey implementation. You can download SML of New Jersey from http://smlnj.org. SML of New Jersey is available for most platforms so you should be able to find an implementation for your needs. You'll want to get the latest working version.

Once you've installed SML you can open a terminal window and start the interpreter. Typing `sml` at the command-line will start the interactive mode of the interpreter. Typing `ctl-d` will terminate the interpreter. You can type expressions and programs directly in at the interpreter's prompt or you can type them in a file and use that file within SML. To do this you type the word *use* as follows:

```
Standard ML of New Jersey v110.59
- use "myfile.txt";
```

SML will take whatever you have typed in the file and evaluate it just as if you had typed it directly into the interpreter.

You should use the examples and practice problems in this chapter to learn SML. The following sections will introduce you to important aspects of SML and will get you ready to write more complicated programs in the next chapter.

5.4 Expressions, Types, Structures, and Functions

Functional programming focuses on the evaluation of expressions. In SML you can evaluate expressions right in the inteperter. When evaluating an expression you will notice that type information is displayed along with the result of the expression evaluation.

Example 5.4

Here are some expression evaluations in SML.

```
- 6;
val it = 6 : int
- 5*3;
val it = 15 : int
- ~1;
val it = ~1 : int
- 5.0 * 3.0;
val it = 15.0 : real
- true;
val it = true : bool
- 5 * 3.0;
stdIn:6.1-6.8 Error: operator and operand don't agree [literal]
  operator domain: int * int
  operand:         int * real
  in expression:
    5 * 3.0
-
```

In SML the identifier *it* is bound to the result of the last successfully evaluated expression. This is convenient if you want to use the result in a subsequent expression. You can refer to the previous result as *it*.

You might notice that a negative one is written as ˜1 in SML. While a little unconventional, ˜ is the unary negation operator in SML, distinguishing it from the binary subtraction operator.

SML has a very rigorous type system. In fact, the type system for SML has been proved sound. That means that any correctly typed program will be free of type errors. In addition, SML is statically typed. That means that all type errors are detected at *compile-time* and not at *run-time*. Robin Milner proved this for Standard ML. ML is the only widely distributed language whose type system has been formally defined[6].

While being formally defined and rigorous, the type system of ML is remarkably flexible. It is polymorphic. We'll see what this means for us soon. Many of the types in ML are also implicitly expressed. You may notice in the example above that the programmer never entered any types for the expressions given there. This frees the programmer from having to key in most of the types in a program since they are mostly determined automatically.

You should have also noticed that there is a type error in example 5.4. ML is polymorphic but it is also strongly typed. Since 5 is an integer in SML and 3.0 is a real, the two cannot be multiplied together. If you should have the need to multiply an integer and a real it can be done, but you must explicitly convert one of the types.

Example 5.5

Here is some code to multiply an integer and a real, producing a real number.

```
- Real.fromInt(5) * 3.0;
[autoloading]
[library $SMLNJ-BASIS/basis.cm is stable]
[autoloading done]
val it = 15.0 : real
-
```

The integer 5 is converted to 5.0 by calling a function called fromInt in the structure called Real. A Structure in SML is a grouping of functions and types. It begins with *structure* name = *struct* and ends with the keyword *end*. Everything after the *struct* and before the *end* is part of the structure. See appendix D for an example of this.

There are several structures that are part of the basis library of SML. The basis library is available in SML when the interpreter is started. Some of these structures are listed in appendix C. The structures in the basis library include Bool, Int, Real, Char, String, and List. By referring to appendix C or looking on the web at http://standardml.org/Basis you can see what functions are included in each of these structures.

A function in SML takes one or more arguments and returns a value. The signature of a function is the type of the function. In other words, a function's type is its signature.

Example 5.6

The signature of the function `fromInt` in the Real structure is

```
val fromInt : int -> real
```

The signature of `fromInt` tells us that it takes an `int` as an argument and returns a `real`. From the name of the function, and the fact that it is part of the Real structure, we can ascertain that it makes a `real` number from an `int`.

The type on the left side of the arrow (i.e. the `->`) is the type of the arguments given to the function. The type on the right side of the arrow is the type of the value returned by the function. The `fromInt` function takes an `int` as an argument and returns a `real`.

☞ Practice 5.4

Write expressions that compute the values described below. Consult the basis library in appendix C as needed.

1. Divide the integer bound to x by 6.
2. Multiply the integer x and the real number y giving the closest integer as the result.
3. Divide the real number 6.3 into the real number bound to x.
4. Compute the remainder of dividing integer x by integer y.

5.5   Recursive Functions

Recursion is the way to get things done in a functional language. Recursion happens when a function calls itself. Because of the principle of referential transparency a function must never call itself with the same arguments. If it were to do that, then the function would do exactly what it did the last time, call itself with the same arguments, which would then.... Well, you get the picture!

To spare ourselves from this problem we insist on two things happening. First, every recursive function must have a base case. A base case is a simple subproblem that we are trying to solve that doesn't require recursion. We must write some code that checks for the simple problem and simply returns the answer in that case.

The second rule of recursive functions requires them to call themselves on some simpler or smaller subproblem. In some way each recursive call should take a step toward the base case of the problem. If each recursive call advances toward the base case then by the mathematical principle of induction we can conclude the function will work for all values on which the function is defined! The trick is not to think about this too hard. The recursive case is often referred to as the inductive case.

To define a function in SML we write the keyword `fun` followed by a function name, parameters, an equal sign, and the body of the function. The syntax is quite similar to defining functions in other languages. The main difference is the body of the function. Instead of being a sequence of statements with variable assignment, the body of the function will be an expression.

One important expression in SML is the *if-then-else* expression. This is not an *if-then-else* statement. Instead, it's an *if-then-else* expression. An *if-then-else* expression gives one of two values and those values must be type compatible. The easiest way to understand *if-then-else* expressions is to see one in practice.

Example 5.7

The Babylonian method of computing square root of a number, x, is to start with an arbitrary number as a *guess*. If $guess^2 = x$ we are done. If not, then let the next guess be $(guess + x/guess)/2.0$. To write this as a recursive function we must find a base case and be certain that our successive guesses will approach the base case. Since the Babylonian method of finding a square root is a well-known algorithm, we can be assured it will converge on the square root. The base case has to be written so that when we get close enough, we will be done. Let's let the *close enough* factor be one millionth of the original number.

The recursive function then looks like this:

```
1   fun babsqrt(x,guess) =
2     if Real.abs(x-guess*guess) < x/1000000.0 then
3       guess
4     else
5       babsqrt(x,(guess + x/guess)/2.0);
```

Looking back at this example there are two things to observe. The base case comes first. If the guess is within one millionth of the right value then the function returns the guess as the square root. The other observation is the recursive call brings us closer to the solution.

☞ Practice 5.5

$n!$ is called the factorial of n. It is defined recursively as $0! = 1$ and $n! = n * (n-1)!$. Write this as a recursive function in SML.

☞ Practice 5.6

The Fibonacci sequence is a sequence of numbers $0, 1, 1, 2, 3, 5, 8, 13, \ldots$. Subsequent numbers in the sequence are derived by adding the previous two numbers in the sequence together. This leads to a recursive definition of the Fibonacci sequence. What is the recursive definition of Fibonacci's sequence? HINT: The first number in the sequence can be thought of as the 0th element, the second the 1st element and so on. So, $fib(0) = 0$. After arriving at the definition, write a recursive SML function to find the nth element of the sequence.

5.6   Characters, Strings, and Lists

SML has separate types for characters and strings. A character literal begins with a pound sign (i.e. #). The character is then surrounded by double quotes. So, the first character in the alphabet is represented as #"a" in SML. There are several functions available in the Char structure for testing and converting characters. The signature of the functions in the Char structure is given in appendix C.4.

Strings in SML are not simply sequences of characters as they are in some languages. A string in SML is its own primitive type. There are functions for converting between strings and characters of course. You can consult appendix C, sections C.4 and C.5 for a list of those functions. A string literal is text surrounded by double quotes. The backslash character (i.e. \) is an escape character in strings. This means to include a double quote in a string you can write \" as part of the string. A \n is the newline character in a string and \t is the tab character as they are in many languages.

Perhaps the most powerful data structure in SML is the list. A list is polymorphic meaning that there are many list types in SML. However, the list functions all work on any type of list. Since it is impossible to determine all the types in SML (because programmers may define their own types), a list's type is parameterized by a type variable. A list's type is written as 'a list. When the type of the list is known, the type variable 'a is replaced by the type it represents. So, a list of integers has type int list. You may have figured this out already, but lists in SML must be homogeneous. This means all the elements of a list must have the same type. This is not like some languages, but there is a good reason for this restriction. Requiring lists to be homogeneous makes static checking of the types in SML possible and the type checker sound and complete.

A list is constructed in one of several ways. First, an empty list is represented as nil or by the empty list (i.e. []). A list may be represented as a literal by putting a left bracket and a right bracket around the list contents, as in [1,4,9,16]. A list may also be constructed using the list constructor which is written ::. The list constructor takes an element on the left side of it and a list on the right side and constructs a new list of its two arguments. A list may be constructed by concatenating two lists together. List concatenation is represented with the @ symbol.

Example 5.8

The following are all valid list constructions in SML.

```
[1,4,9,16]
1::[4,9,16,25]
#"a"::#"b"::[#"c"]
1::2::3::nil
["hello","how"]@["are","you"]
```

The third example works because the :: constructor is right-associative. So the right-most constructor is applied first, then the one to its left, and so on.

Example 5.9

The signatures of the list constructor and some list functions are given here.

```
:: : 'a * 'a list -> 'a list
@ : 'a list * 'a list -> 'a list
hd : 'a list -> 'a
tl : 'a list -> 'a list
```

☞ Practice 5.7

The following are NOT valid list constructions in SML. Why not? Can you fix them?

```
#"a"::["beautiful day"]
"hi"::"there"
["how","are"]::"you"
[1,2.0,3.5,4.2]
2@[3,4]
[]::3
```

You can select elements from a list using the hd and tl functions. The hd (pronounced *head*) of a list is the first element of the list. The tl is the tail or all the rest of the elements of the list. Calling the hd or tl functions on the empty list will result in an error. Using these two functions and recursion it is possible to access each element of a list.

Example 5.10

Here is a function called implode that takes a list of characters as an argument and returns a string comprised of those characters.

```
fun implode(lst) =
  if lst = [] then ""
  else str(hd(lst))^implode(tl(lst))
```

So, implode([#"H",#"e",#"l",#"l",#"o"]) would yield "Hello".

When writing a recursive function the trick is to not think too hard about how it works. Think of the base case or cases and the recursive cases separately. So, in the function above the base case is when the list is empty (since a list is the parameter). When the list is empty, the string the function should return would also be empty, right?

The recursive case is when when the list is not empty. In that case, there is at least one element in the list. If that is true then we can call hd to get the first element and tl to get the rest of the list. The head of the list is a character and must be converted to a string. The rest of the list is converted to a string by calling some function that will convert a list to a string. This function is called implode! We can just assume it will work. That is the nature of recursion. The trick, if there is one, is to trust that recursion will work. Later, we will explore exactly why we can trust recursion.

☞ Practice 5.8

Write a function called `explode` that will take a string as an argument and return a list of characters in the string. So, `explode("hi")` would yield `[#"h",#"i"]`. HINT: How do you get the first character of a string?

Example 5.11

Here are a couple more examples of list functions.

```
1  fun length(x) =
2      if null x then 0
3      else 1+length(tl(x))
4  fun append(L1, L2) =
5      if null L1 then L2 else hd(L1)::append(tl(L1),L2)
```

☞ Practice 5.9

Use the append function to write reverse. The reverse function reverses the elements of a list. Its signature is

```
reverse = fn: 'a list -> 'a list
```

5.7 Pattern Matching

Frequently, recursive functions rely on several recursive and several base cases. SML includes a nice facility for handling these different cases in a recursive definition by allowing pattern matching of the arguments to a function. Pattern matching works with literal values like 0, the empty string, and the empty list. Generally, you can use pattern matching if you would normally use equality to compare values. Real numbers are not equality types. The *real* type only approximates real numbers. Example 5.7 shows how to compare real numbers for equality.

You can also use constructors in patterns. So the list constructor `::` works in patterns as well. Functions like the append function (i.e. @) and string concatenation (i.e. ^) don't work in patterns. These functions are not constructors of values and cannot be efficiently or deterministically matched to patterns of arguments.

Example 5.12

Append can be written using pattern-matching as follows. The extra parens around the recursive call to append are needed because the `::` constructor has higher precedence than function application.

```
fun append(nil,L2) = L2
  | append(h::t,L2) = h::(append(t,L2))
```

☞ Practice 5.10

Rewrite reverse using pattern-matching.

5.8 Tuples

A tuple type is a cross product of types. A two-tuple is a cross product of two types, a three-tuple is a cross product of three types, and so on.

Example 5.13

(5,6) is a two-tuple of int * int.
The three tuple (5,6,"hi") is of type int * int * string.

You might have noticed the signature of some of the functions in this chapter. For instance, consider the signature of the append function. Its signature is

```
val append : 'a list * 'a list -> 'a list
```

This indicates it's a function that takes as its argument an 'a list * 'a list tuple. In fact, every function takes a single argument and returns a single value. The sole argument might be a tuple of one or more values, but every function takes a single argument as a parameter. The return value of a function may also be a tuple.

In many other languages we think of writing function application as the function followed by a left paren, followed by comma separated arguments, followed by a right paren. In Standard ML (and most functional languages) function application is written as a function name followed by the value to which the function is applied. This is just like function application in the lambda calculus. So, we can think of calling a function with zero or more values, but in reality we are passing one value to a every function in ML which may be a tuple.

Example 5.14

In Standard ML rather than writing

```
append([1,2],[3])
```

it is more appropriate to write

```
append ([1,2],[3])
```

because function application is a function name followed by the value it will be applied to. In this case append is applied to a tuple of 'a list * 'a list.

5.9 Let Expressions and Scope

Let expressions are simply syntax for binding a value to an identifier to later be used in an expression. They are useful when you want to document your code by assigning a meaningful name to a value. They can also be useful when you need the same value more than once in a function definition. Rather than calling a function twice to get the same value, you can call it once and bind the value to an identifier. Then the identifier can be used as many times as the value is needed. This is more efficient than calling a function multiple times with the same arguments.

Example 5.15

Consider a function that computes the sum of the first n integers.

```
1  fun sumupto(0) = 0
2    | sumupto(n) =
3      let val sum = sumupto(n-1)
4      in
5          n + sum
6      end
```

Let expressions let us define identifiers that are local to functions. The identifier called sum in the example above is not visible outside the sumupto function definition. We say the scope of sum is the body of the let expression (i.e. the expression given between the in and end keywords). Let expressions allow us to declare identifiers with limited scope.

Limiting scope is an important aspect of any language. Function definitions also limit scope in SML and most languages. The formal parameters of a function definition are not visible beyond the body of the function.

Binding values to identifiers should not be confused with variable assignment. A binding of a value to an identifier is a one time operation. The identifier's value cannot be updated like a variable. A practice problem will help to illustrate this.

☞ Practice 5.11

What is the value of x at the various numbered points within the following expression? Be careful, it's not what you think it might be if you are relying on your imperative understanding of code.

```
1  let val x = 10
2  in
3      (* 1. Value of x here? *)
4      let val x = x+1
5      in
6          (* 2. Value of x here? *)
7          x
8      end;
9      (* 3. Value of x here? *)
10      x
11  end
```

Bindings are not the same as variables. Bindings are made once and only once and cannot be updated. Variables are meant to be updated as code progresses. Bindings are an association between a value and an identifier that is not updated.

SML and many modern languages use static or lexical scope rules. This means you can determine the scope of a variable by looking at the structure of the program without considering its execution. The word lexical refers to the written word and lexical or static scope refers to determining scope by looking at how the code is written and not the execution of the code. Originally, LISP used dynamic scope rules. To determine dynamic scope you must look at the bindings that were active when the code being executed was called. The difference between dynamic and static scope can be seen when functions may be nested in a language and may also be passed as parameters or returned as function results.

Example 5.16

The difference between dynamic and static scope can be observed in the following program.

```
1   let fun a () =
2             let val x = 1
3                 fun b () = x
4             in
5                 b
6             end
7         val x = 2
8         val c = a ()
9   in
10      c ()
11  end
```

In this program the function *a*, when called, declares a local binding of *x* to 1 and returns the function *b*. When *c*, the result of calling *a*, is called it returns a 1, the value of *x* in the environment where *b* was defined, not a 2. This result is what most people expect to happen. It is static or lexical scope. The correct value of *x* does not depend on the value of *x* when it was called, but the value where the function *b* was written.

While static scope is used by many programming languages including Standard ML, Python, Lisp, and Scheme, it is not used by all languages. The Emacs version of Lisp uses dynamic scope and if the equivalent Lisp program is evaluated in Emacs Lisp it will return a value of 2 from the code in example 5.16.

It is actually harder to implement static scope than dynamic scope. In dynamically scoped languages when a function is returned as a value the return value can include a pointer to the code of the function. When the function *b* from example 5.16 is executed in a dynamically scoped language, it simply looks in the current environment for the value of *x*. To implement static scope, more than a pointer to the code is needed. A pointer to the current environment is needed which contains the binding of *x* to the value at the time the function was defined. This is needed so when the function *b* is evaluated, the right *x* binding can be found. The combination of a pointer to a function's code and its environment is called a *closure*. Closures are used to represent function values in statically scoped languages where functions may be returned as results and nested functions may be defined.

5.10 Datatypes

The word datatype is often loosely used in computer science. In ML, a datatype is a special kind of type. A datatype is a tagged structure that can be recursively defined. This type is powerful in that you can define enumerated types with it and you can define recursive data structures like lists and trees.

Datatypes are user-defined types and are generally recursively defined so there are infinitely many datatypes in Standard ML. This would be something like creating a class in C++ except that classes define both data and methods. In a functional language a set of functions is defined to work with a type of data through pattern-matching as described in section 5.7.

Example 5.17

In C/C++ we can create an enumerated type by writing

```
1  enum TokenType {
2      identifier,keyword,number,add,sub,times,divide,lparen,
3      rparen,eof,unrecognized
4  };
```

This defines a type called TokenType of eleven values: identifier==0, keyword==1, number==2, etc. You can declare a variable of this type as follows:

```
TokenType t = keyword;
```

However, there is nothing preventing you from executing the statement

```
t = 1; //this is the keyword value.
```

In this example, even though t is of type TokenType, it can be assigned an integer. This is because the TokenType type is just another name for the integer type in C++. Assigning t to 1 doesn't bother C++ in the least. In fact, assigning t to 99 wouldn't bother C++ either. In ML, we can't use integers and datatypes interchangeably.

```
- datatype TokenType = Identifier | Keyword | Number |
     Add | Sub | Times | Divide | LParen | RParen | EOF |
     Unrecognized;
datatype TokenType = Identifier | Keyword | Number | ...
- val x = Keyword;
x = Keyword : TokenType
```

Datatypes allow programmers to define their own types. Normally, a datatype includes other information. They are used to represent structured data of some sort.

By adding the keyword `of`, a datatype value can include a tuple of other types as part of its definition. A datatype can represent any kind of recursive data structure. That includes lists, trees, and other structures that are related to lists and trees. In the example below a tree definition with a mix of unary and binary nodes is defined.

Datatypes allow a programmer to write a recursive function that can traverse the data given to it. Functions can use pattern matching to handle each case in a datatype with a pattern match in the function.

Example 5.18

In this datatype the `add'` value can be thought of as a node in an AST that has two children, each of which are ASTs. The datatype is recursive because it is defined in terms of itself.

```
1  datatype
2    AST = add' of AST * AST
3        | sub' of AST * AST
4        | prod' of AST * AST
5        | div' of AST * AST
6        | negate' of AST
7        | integer' of int
8        | store' of AST
9        | recall';
```

Example 5.18 is the entire definition of abstract syntax trees for expressions in the calculator language. In addition to the nodes you've seen before, the negate' node represents unary negation of the value we get when evaluating its child. So now -6 is a valid expression.

Example 5.19

The abstract syntax tree for -6S+R would be as shown below.

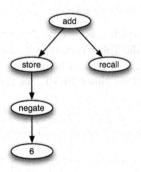

`add'(store'(negate'(integer'(6)))), recall')` is the value of the tree given above as an SML datatype. A function can be written to evaluate an abstract synatx tree based on the patterns in a value like this.

You can use pattern matching on datatypes. For instance, to evaluate an expression tree you can write a recursive function using pattern-matching. Each pattern that is matched in such a function corresponds to processing one node in the tree. Each subtree can be processed by a recursive call to the same function. In the function below, the parameter `min` is the value of the memory before evaluating the given node in the abstract syntax tree. The value `mout` is the value of memory after evaluating the node in the abstract syntax tree.

Example 5.20

This example illustrates how to use pattern-matching with datatypes and patterns in a `let` construct.

```
1    fun evaluate(add'(e1,e2),min) =
2        let val (r1,mout1)= evaluate(e1,min)
3            val (r2,mout) = evaluate(e2,mout1)
4        in
5            (r1+r2,mout)
6        end
7
8      | evaluate(sub'(e1,e2),min) =
9        let val (r1,mout1)= evaluate(e1,min)
10           val (r2,mout) = evaluate(e2,mout1)
11       in
12           (r1-r2,mout)
13       end
```

`mout1` is the value of memory after evaluating `e1`. This is passed to evaluating `e2` as the value of the memory before evaluating `e2`. The value of memory after evaluating `e2` is the value of memory after evaluating the sum/difference of the two expressions. This pattern of passing the memory through the evaluation of the tree is called *single-threading* the memory in the computation.

☞ Practice 5.12

Define a datatype for integer lists. A list is constructed of a head and a tail. Sometimes this constructor is called `cons`. The empty list is also a list and is usually called `nil`. However, in this practice problem , to distinguish from the built-in `nil` you could call it `nil'`.

☞ Practice 5.13

Write a function called `maxIntList` that returns the maximum integer found in one of the lists you just defined in practice problem 5.12. You can consult appendix C for help with finding the max of two integers.

5.11 Parameter Passing in Standard ML

The types of data in Standard ML include integers, reals, characters, strings, tuples, lists, and the user-defined datatypes presented in section 5.10. If you look at these types in this chapter and in appendix C you may notice that there are no functions that modify the existing data. The substring function defined on strings returns a new string as do all functions on the types of data available in Standard ML. All data in Standard ML is immutable. That's quite a statement. It's true of some functional languages, but not all. Every type of data in Standard ML is immutable.

Well, almost. There is one type of data that is mutable in ML. A reference is a reference to a value of a determined type. References may be mutated to enable the programmer to program using the imperative style of programming. References are discussed in more detail in section 5.19.

The absence of mutable data, except for references, has some impact on the implementation of the language. Values are passed by reference in Standard ML. However, the only time that matters is when a reference is passed as a parameter. Otherwise, the immutability of all data means that how data is passed to a function is irrelevant. This is nice for programmers as they don't have to be concerned about which functions mutate data and which construct new data values. There is only one operation that mutates data, the assignment operator (i.e. :=) as described in section 5.19 and the only data it can mutate is a reference. In addition, because all data is immutable and passed by reference, parameters are passed efficiently in ML like constant references of C++.

5.12 Efficiency of Recursion

Once you get used to it, writing recursive functions isn't too hard. In fact, it can be easier than writing iterative solutions. But, just because you find a recursive solution to a problem, doesn't mean it's an effficient solution to a problem. Consider the Fibonacci numbers. The recursive definition leads to a very straightforward recursive solution. However, as it turns out, the simple recursive solution is anything but efficient. In fact, given the definition in example 5.21, fib(43) took twenty-four seconds to compute on a 1.5 GHz G4 PowerBook with 1GB of RAM.

Example 5.21

The Fibonacci numbers can be computed with the function definition given below.

```
fun fib(0) = 0
  | fib(1) = 1
  | fib(n) = fib(n-1) + fib(n-2)
```

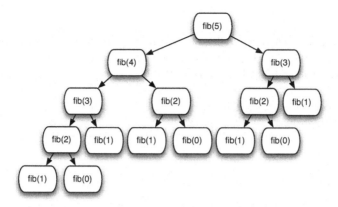

Fig. 5.1: Calls to calculate fib(5)

This is a very inefficient way of calculating the Fibonacci numbers. The number of calls to fib increases exponentially with the size of n. This can be seen by looking at a tree of the calls to fib as in figure 5.1. The number of calls required to calculate `fib(5)` is 15. If we were to enumerate the calls required to calculate `fib(6)` it would be everything in the `fib(5)` call tree plus the number of nodes in the `fib(4)` call tree, $15 + 9 = 25$. The number of calls grows exponentially.

☞ Practice 5.14

One way of proving that the `fib` function given above is exponential is to show that the number of calls for `fib(n)` is bounded by two exponential functions. In other words, there is an exponential function of n that will always return less than the number of calls required to compute `fib(n)` and there is another exponential function that always returns greater than the number of required calls to compute `fib(n)` for some choice of starting n and all values greater than it. If the number of calls to compute `fib(n)` lies in between then the `fib` function must have exponential complexity. Find two exponential functions of the form c^m that bound the number of calls required to compute `fib(n)`.

From this analysis you have probably noticed that there is a lot of the same work being done over and over again. It may be possible to eliminate a lot of this work if we are smarter about the way we write the Fibonacci function. In fact it is. The key to this efficient version of `fib` is to recognize that we can get the next value in the sequence by adding together the previous two values. If we just carry along two values, the current and the next value in the sequence, we can compute each Fibonacci number with just one call. Example 5.22 shows you how to do this. With the new function, computation of fib(43) is instantaneous.

Example 5.22

Using a helper function may lead to a better implementation in some situations. In the case of the `fib` function, the `fibhelper` function turns an exponentially complex function into a linear time function.

```
1  fun fib(n) =
2     let fun fibhelper(count,current,previous) =
3             if count = n then previous
4             else fibhelper(count+1,previous+current,current)
5     in
6             fibhelper(0,1,0)
7     end
```

In example 5.22 we used a helper function that was private to the `fib` function because we don't want other programmers to call the `fibhelper` function directly. It is meant to be used by the `fib` function. We also wouldn't want to have to remember how to call the `fibhelper` function each time we called it. By hiding it in the `fib` function we can expose the same interface we had with the original implementation, but implement a much more efficient function.

The helper function uses a pattern called an accumulator pattern. The helper function makes use of an accumulator to reduce the amount of work that is done. The work is reduced because the function keeps track of the last two values computed by the helper function to aide in computing the next number.

☞ Practice 5.15

Consider the reverse function you wrote in practice problem 5.9. The `append` function is called n times, where n is the length of the list. How many cons operations happen each time append is called? What is the overall complexity of the reverse function?

5.13 Tail Recursion

One criticism of functional programming centers on the heavy use of recursion that is seen by some critics as overly inefficient. The problem is related to the use of caches in modern processors. Depending on the block size of an instruction cache, the code surrounding the currently executing code may be readily available in the cache. However, when the instruction stream is interrupted by a call to a function, even the same function, the cache may not contain the correct instructions. Retrieving instructions from memory is much slower than finding them in the cache. However, cache sizes continue to increase and even imperative languages like C++ and Java encourage many calls to small functions or methods given their object-oriented nature. So, the argument in favor of fewer function calls has certainly diminished in recent years.

It is still the case that a function call takes longer than executing a simple loop. When a function call is made, extra instructions are executed to create a new activation record. In addition, in pipelined processors the pipeline is disrupted by function calls. Standard ML of New Jersey, Scheme, and some other functional languages have a mechanism where they optimize certain recursive functions by reducing the storage on the run-time stack and eliminating calls. In certain cases, recursive calls can be automatically transformed to code that can be executed using jump or branch instructions. For this optimization to be possible, the recursive function must be tail recursive. A tail recursive function is a function where the very last operation of the function is the recursive call to itself.

Example 5.23

This is the `factorial` function.

```
fun factorial 0 = 1
  | factorial n = n * factorial (n-1);
```

Is factorial tail recursive? The answer is no. Tail recursion happens when the very last thing done in a recursive function is a call to itself. The last thing done above is the multiplication.

When factorial 6 is invoked, activation records are needed for seven invocations of the function, namely factorial 6 through factorial 0. Without each of these stack frames, the local values of n, n=6 through n=0, will be lost so that the multiplication at the end can not be carried out correctly.

At its deepest level of recursion all the information in the expression,

$$(6 * (5 * (4 * (3 * (2 * (1 * (factorial\ 0)))))))$$

is stored in the run-time execution stack.

Practice 5.16

Show the run-time execution stack at the point that factorial 0 is executing when the original call was factorial 6.

The `factorial` function can be written to be tail recursive. The solution is to use a technique similar to the `fib` function improvement made in example 5.22. An accumulator is added to the function definition. An accumulator is an extra parameter that can be used to accumulate a value, much the way you would accumulate a value in a loop. The accumulator value is initially given the identity of the operation used to accumulate the value. In example 5.24 the operation is multiplication. The identity provided as the initial value is 1.

This is the tail recursive version of the `factorial` function. The tail recursive function is the `tailfac` helper function.

```
1  fun factorial n =
2     let fun tailfac(0,prod) = prod
3         | tailfac(n,prod) = tailfac(n-1,prod*n)
4     in
5        tailfac(n,1)
6     end
```

Note that although `tailfac` is recursive, there is no need to save it's local environment when it calls itself since no computation remains after the call. The result of the recursive call is simply passed on as the result of the current function call. A function is tail recursive if its recursive call is the last action that occurs during any particular invocation of the function.

☞ Practice 5.17

Use the accumulator pattern to devise a more efficient reverse function. The append function is not used in the efficient reverse function. HINT: What are we trying to accumulate? What is the identity of that operation?

5.14 Currying

A binary function, for example, + or @, takes both of its arguments at the same time. a+b will evaluate both a and b so that values can be passed to the addition operation. There can be an advantage in having a binary function take its arguments one at a time. Such a function is called *curried* after Haskell Curry. In fact, ML functions can take their parameters one at a time.

The preceding paragraph may be a bit misleading. Every ML function takes one and only one parameter. So a curried function takes one argument as well. However, that function of one parameter may in turn return a function that takes a single argument. This is probably best illustrated with an example.

Here is a function that takes a pair of arguments as its input.

```
- fun plus(a:int,b) = a+b;
val plus = fn : int * int -> int
```

The function `plus` takes one argument that just happens to be a tuple. It is applied to a single tuple.

```
- plus (5,8);
val it = 13 : int
```

ML functions can be defined with what looks like more than one parameter:

```
fun cplus(a:int) b = a+b;
val cplus = fn : int -> (int -> int )
```

Observe the signature of the function `cplus`. It takes two arguments, but takes them one at a time. Actually, cplus takes only one argument. The `cplus` function returns a function that takes the second argument. The second function has no name.

```
cplus 5 8;
val it = 13 : int
```

Function application is left associative. The parens below show the order of operations.

```
(cplus 5) 8;
val it = 13 : int
```

The result of `(cplus 5)` is a function that adds 5 to its argument.

```
- cplus 5;
val it = fn : int -> int
```

We can give this function a name.

```
- val add5 = cplus 5;
val add5 = fn : int -> int
- add5 8;
val it = 13 : int
```

☞ Practice 5.18

Write a function that given an uncurried function of two arguments will return a curried form of the function so that it takes its arguments one at a time.
Write a function that given a curried function that takes two arguments one at a time will return an uncurried version of the given function.

Curried functions allow partial evaluation, a very interesting topic in functional languages, but beyond the scope of this text. It should be noted that Standard ML of New Jersey uses curried functions extensively in its implementation.

5.15 Anonymous Functions

Section 5.2 describes the lambda calculus. In that section we learned that functions can be characterized as first class objects. Functions can be represented by a lambda abstraction and don't have to be assigned a name. This is also true in SML. Functions in SML don't need names.

Example 5.26

The anonymous function $\lambda x, y.y^2 + x$ can be represented in ML as

```
- fn x => fn y => y*y + x;
```

The anonymous function can be applied to a value in the same way a named function is applied to a value. Function application is always the function first, followed by the value.

```
- (fn x => fn y => y*y + x) 3 4;
val it = 19 : int
```

We can define a function by binding a lambda abstraction to an identifier:

```
- val f = fn x => fn y => y*y + x;
val f = fn: int -> int -> int
- f 3 4;
val it = 19 : int
```

This mechanism provides an alternative form for defining functions as long as they are not recursive; in a `val` declaration, the identifier being defined is not visible in the expression on the right side of the arrow. For recursive definitions, use `val rec`.

Example 5.27

To define a recursive function using the anonymous function form you must use `val rec` to declare it.

```
- val rec fac = fn n => if n=0 then 1 else n*fac(n-1);
val fac = fn: int -> int
- fac 7;
val it = 5040:int
```

The form of function definition presented in example 5.27 is the way all functions are defined in SML. The functional form used when the keyword `fun` is used to define a function is translated into the form show in example 5.27. The `fun` form of function definition is *syntactic sugar*. Syntactic sugar refers to another way of writing something that gets treated the same way in either case.

5.16 Higher-Order Functions

The unique feature of functional languages is that functions are treated as first-class objects with the same rights as other objects, namely to be stored in data structures, to be passed as a parameter, and to be returned as function results. Functions can be bound to identifiers using the keywords `fun`, `val`, and `val rec` and may also be stored in structures:

Example 5.28

These are examples of functions being treated as values.

```
- val fnlist = [fn (n) => 2*n, abs, ~, fn (n) => n*n];
val fnlist = [fn,fn,fn,fn] : (int -> int) list
```

Notice each of these functions takes an int and returns an int. An ML function can be defined to apply each of these functions to a number.

Example 5.29

The construction function applies a list of functions to a value.

```
fun construction  nil n = nil
  | construction (h::t) n = (h n)::(construction t n);
val construction = fn : ('a -> 'b) list -> 'a -> 'b list

construction [op +, op *, fn (x,y) => x - y] (4,5);
val it = [9,20,~1] : int list
```

Construction is based on a functional form found in FP, a functional language developed by John Backus. It illustrates the possibility of passing functions as arguments. Since functions are first-class objects in ML, they may be stored in any sort of structure. It is possible to imagine an application for a stack of functions or even a tree of functions.

A function is called higher-order if it takes a function as a parameter or returns a function as its result. Higher-order functions are sometimes called functional forms since they allow the construction of new functions from already defined functions.

The usefulness of functional programming comes from the use of functional forms that allow the development of complex functions from simple functions using abstract patterns. The construction function is one of these abstract patterns of computation. These functional forms, or patterns of computation, appear over and over again in programs. Programmers have recognized these patterns and have abstracted out the details to arrive at several commonly used higher-order functions. The next sections introduce you to several of these higher-order functions.

Composition

Composing two functions is a naturally higher-order operation that you have probably used in algrebra. Have you ever written something like f(g(x))? This operation can be expressed in ML. In fact, ML has a built-in operator called o which represents composition. Example 5.30 shows you how composition can be written and used.

Example 5.30

```
- fun compose f g x = f (g x);
val compose = fn : ('a -> 'b) -> ('c -> 'a) -> 'c -> 'b
- fun add1 n = n+1;
- fun sqr n:int = n*n;
- val incsqr = compose add1 sqr;
val incsqr = fn : int -> int
- val sqrinc = compose sqr add1;
val sqrinc = fn : int -> int
```

Observe that these two functions, incsqr and sqrinc, are defined without the use of parameters.

```
- incsqr 5;
val it = 26 : int
- sqrinc 5;
val it = 36 : int
```

ML has a predefined infix function o that composes functions. Note that o is uncurried.

```
- op o;
val it = fn : ('a -> 'b) * ('c -> 'a) -> 'c -> 'b
- val incsqr = add1 o sqr;
val incsqr = fn : int -> int
- incsqr 5;
val it = 26 : int
- val sqrinc = op o(sqr,add1);
val sqrinc = fn : int -> int
- sqrinc 5;
val it = 36 : int
```

Map

In SML, applying a function to every element in a list is called map and is predefined. It takes a unary function and a list as arguments and applies the function to each element of the list returning the list of results.

Example 5.31

```
- map;
val it = fn : ('a -> 'b) -> 'a list -> 'b list
- map add1 [1,2,3];
val it = [2,3,4] : int list
- map (fn n => n*n - 1) [1,2,3,4,5];
```

```
val it = [0,3,8,15,24] : int list
- map (fn ls => "a"::ls) [["a","b"],["c"],["d","e","f"]];
val it = [["a","a","b"],["a","c"],["a","d","e","f"]] :
             string list list
- map real [1,2,3,4,5];
val it = [1.0,2.0,3.0,4.0,5.0] : real list
```

map can be defined as follows:

```
fun map f nil = nil
  | map f (h::t) = (f h)::(map f t);
```

☞ Practice 5.19

Describe the behavior (signatures and output) of these functions:

```
map (map add1)
(map map)
```

Invoking (map map) causes the type inference system of SML to report

```
stdIn:12.27-13.7 Warning: type vars not generalized
   because of value restriction are instantiated to
   dummy types (X1,X2,...)
```

This warning message is OK. It is telling you that to complete the type inference for this expression, SML had to instantiate a type variable to a dummy variable. When more type information is available, SML would not need to do this. The warning message only applies to the specific case where you created a function by invoking (map map). In the presence of more information the type inference system will interpret the type correctly without any dummy variables.

Reduce or Foldright

Higher-order functions are developed by abstracting common patterns from programs. For example, consider the functions that find the sum or the product of a list of integers. In this pattern the results of the previous invocation of the function are used in a binary operation with the next value to be used in the computation.

In other words, to add up a list of values you start with either the first or last element of the list and then add it together with the value next to it. Then you add the result of that computation to the next value in the list and so on. When we start with the end of the list and work our way backwards through the list the operation is sometimes called foldr (i.e. foldright) or reduce.

Example 5.32

```
fun sum nil = 0
  | sum ((h:int)::t) = h + sum t;

val sum = fn : int list -> int
sum [1,2,3,4,5];
val it = 15 : int

fun product nil = 1
  | product ((h:int)::t) = h * product t;

val product = fn : int list -> int
product [1,2,3,4,5];
val it = 120 : int
```

Each of these functions has the same pattern. If we abstract the common pattern as a higher-order function we arrive at a common higher-order function called `foldr` or `reduce`. `foldr` is an abbreviation for foldright.

Example 5.33

This function is sometimes called `foldr`. In this example it is called `reduce`.

```
fun reduce f init nil = init
  | reduce f init (h::t) = f(h, reduce f init t);

val reduce = fn : ('a * 'b -> 'b) -> 'b -> 'a list -> 'b
reduce op + 0 [1,2,3,4,5];
val it = 15 : int
reduce op * 1 [1,2,3,4,5];
val it = 120 : int
```

Now `sum` and `product` can be defined in terms of `reduce`.

```
val sumlist = reduce (op +) 0;
val sumlist = fn : int list -> int
val mullist = reduce op * 1;
val mullist = fn : int list -> int
sumlist [1,2,3,4,5];
val it = 15 : int
mullist [1,2,3,4,5];
val it = 120 : int
```

SML includes two predefined functions that reduce a list, `foldr` and `foldl` which stands for foldleft. They behave slightly differently.

Example 5.34

```
foldr;
val it = fn : ('a * 'b -> 'b) -> 'b -> 'a list -> 'b
foldl;
val it = fn : ('a * 'b -> 'b) -> 'b -> 'a list -> 'b
- fun abdiff (m,n:int) = abs(m-n);
val abdiff = fn : int * int -> int
- foldr abdiff 0 [1,2,3,4,5];
val it = 1 : int
- foldl abdiff 0 [1,2,3,4,5];
val it = 3 : int
```

☞ Practice 5.20

How does `foldl` differ from `foldr`? Determine the difference by looking at the example above. Then, describe the result of these functions invocations.

```
foldr op :: nil ls
foldr op @ nil ls
```

Filter

A predicate function is a function that takes a value and returns true or false depending on the value. By passing a predicate function, it is possible to filter in only those elements from a list that satisfy the predicate. This is a commonly used higher-order function called `filter`.

Example 5.35

If we had to write filter ourselves, this is how it would be written. This example also shows how it might be used.

```
fun filter bfun nil = nil
  | filter bfun (h::t) = if bfun h then h::filter bfun t
                            else filter bfun t;

val it = fn : ('a -> bool) -> 'a list -> 'a list
even;
val it = fn : int -> bool
filter even [1,2,3,4,5,6];
val it = [2,4,6] : int list
filter (fn n => n > 3) [1,2,3,4,5,6];
val it = [4,5,6] : int list
```

☞ Practice 5.21

Use filter to select numbers from a list that are

1. divisible by 7
2. greater than 10 or equal to zero

5.17 Continuation Passing Style

Continuation Passing Style (or CPS) is a way of writing functional programs where control is made explicit. In other words, the continuation represents the remaining work to be done. This style of writing code is interesting because the style is used in the SML compiler.

Example 5.36

To understand cps it's best to look at an example. Let's consider the len function for computing the length of a list.

```
fun len nil = 0
  | len (h::t) = 1+(len t);
```

To transform this to cps form we represent the rest of the computation explicitly as a parameter called k. In this way, whenever we need the continuation of the calculation, we can just write the identifier k. Here's the cps form of len.

```
fun cpslen nil k = k 0
  | cpslen (h::t) k = cpslen t (fn v => (k (1 + v)));
```

And here's how cpslen would be called.

```
cpslen [1,2,3] (fn v => v);
```

☞ Practice 5.22

Trace the execution of cpslen to see how it works and how the continuation is used.

Notice that the recursive call to cpslen is the last thing that is done. This function is tail recursive. However, tail recursion elimination cannot be applied because the function returns a function and recursively calls itself with a function as a parameter. CPS is still important because it can be optimized by a compiler. In addition, since control flow is explicit (passed around as k), function calls can be implemented with jumps and many of the jumps can be eliminated if the code is organized in the right way.

Eliminating calls and jumps is important since calls have the effect of interrupting pipelines in RISC processors. Since functional languages make lots of calls, one of the criticisms of functional languages is that they were inefficient. With the optimization of CPS functions, functional languages get closer to being as efficient as imperative languages. In addition, as cache sizes and processor speeds increase the performance difference becomes less and less of an issue.

☞ Practice 5.23

Write a function called `depth` that prints the longest path in a binary tree. First create the datatype for a binary tree. You can use the `Int.max` function in your solution, which returns the maximum of two integers.
First write a non-cps `depth` function, then write a cps `cpsdepth` function.

5.18   Input and Output

SML contains a TextIO structure as part of the basis library. The signature of the functions in the TextIO structure is given in appendix C.7. It is possible to read and write strings to streams using this library of functions. The usual standard input, standard output, and standard error streams are predefined.

Example 5.37

Here is an example of reading a string from the keyboard. Explode is used on the string to show the vector type is really the string type. It also shows how to print something to a stream.

```
- val s = TextIO.input(TextIO.stdIn);
hi there
val s = "hi there\n" : vector
- explode(s);
val it = [#"h",#"i",#" ",#"t",#"h",#"e",
          #"r",#"e",#"\n"] : char list
- TextIO.output(TextIO.stdOut,s^"How are you!\n");
hi there
How are you!
val it = () : unit
```

Since streams can be directed to files, the screen, or across the network, there really isn't much more to input and output in SML. Of course if you are opening your own stream it should be closed when you are done with it. Program termination will also close any open streams.

There are some TextIO functions that may or may not return a value. In these cases an `option` is returned. An `option` is a value that is either NONE or SOME *value*. An option is SML's way of dealing with functions that may or may not succeed. Functions must always return a value or end with an exception. To prevent the exception handling mechanism from being used for input operations that may or may not succeed, this idea of an option was created. Options fit nicely into the strong typing the SML provides.

Example 5.38

The `input1` function of the TextIO structure reads exactly one character from the input and returns an `option` as a result. The reason it returns an `option` and not the character directly is because the stream might not be ready for reading. The `valOf` function can be used to get the value of an `option` that is not NONE.

```
- val u = TextIO.input1(TextIO.stdIn);
hi there
val u = SOME #"h" : elem option
- =
= ^C
Interrupt
- u;
val it = SOME #"h" : elem option
- val v = valOf(u);
val v = #"h" : elem
```

5.19 Programming with Side-effects

Standard ML is not a pure functional language. It is possible to write programs with side effects, such as reading from and writing to streams. To write imperative programs the language should support sequential execution, variables, and possibly loops. All three of these features are available in SML. The following sections show you how to use each of these features.

Variable Declarations

There is only one kind of variable in Standard ML. Variables are called references. It is interesting to note that you cannot update an integer, real, string, or any other type of value in SML. All values are immutable. They cannot be changed once created. That is a nice feature of a language because then you don't have to worry about the distinction between a reference to a value and the value itself.

A reference in Standard ML is typed. It is either a reference to an *int*, or a *string*, or some other type of data. References can be mutated. So a reference can be updated to point to a new value as your program executes. Declaring and using a reference variable is described in the example below.

Example 5.39

In SML a variable is declared by creating a reference to a value of a particular type.

```
- val x = ref 0;
val x = ref 0 : int ref
```

The exclamation point is used to refer to the value a reference points to. This is called the dereference operator. It is the similar to the star (i.e. *) in C++.

```
- !x;
val it = 0 : int
- x := !x + 1;
val it = () : unit
- !x;
val it = 1 : int
```

The assignment operator (i.e. :=) operator updates the reference variable to point to a new value. The result of assignment is the empty tuple which has a special type called unit. Imperative programming in SML will often result in the unit type. Unlike ordinary identifiers you can bind to values using a let val *id* = *Expr* in *Expr* end, a reference can truly be updated to point to a new value.

it should be noted that references in Standard ML are typed. When a reference is created it can only point to a value of the same type it was originally created to refer to. This is unlike references in Python, but is similar to references in Java. A reference refers to a particular type of data.

Sequential Execution

If a program is going to assign variables new values or read from and write to streams it must be able to execute statements or expressions sequentially. There are two ways to write a sequence of expressions in SML. When you write a let val *id* = *Expr* in *Expr* end expression, the *Expr* in between the in and end may be a sequence of expressions. A sequence of expressions is semicolon separated.

Example 5.40

This demonstrates how to write a sequence of expressions.

```
let val x = ref 0
in
  x:= !x + 1;
  TextIO.output(TextIO.stdOut,"The new value of x is "^
                Int.toString(!x)^"\n");
  !x
end
```

Evaluating this expression produces the following output.

```
The new value of x is 1
val it = 1 : int
```

In example 5.40 semicolons separate the expressions in the sequence. Notice that semicolons don't terminate each line as in C++ or Java. Semicolons in SML are expression separators, not statement terminators. The last expression in a sequence of expressions is the return value of the expression. The ! x is the last expression in the sequence above so 1 is returned as the value.

There are times when you may wish to evaluate a sequence of expressions in the absence of a let expression. In that case the sequence of expressions may be surrounded by parens. A left paren can start a sequence of expressions terminated by a right paren. The sequence of expressions is semicolon separated in either case.

Example 5.41

Here is some code that prints the value of x to the screen and then returns x + 1.

```
(TextIO.output(TextIO.stdOut,"The value of x is " ^
Int.toString(x);
x+1)
```

Iteration

Strictly speaking, variables and iteration are not needed in a functional language. Parameters can be passed in place of variable declarations. Recursion can be used in place of iteration. However, there are times when an iterative function might make more sense. For instance, when reading from a stream it might be more efficient to read the stream in a loop, especially when the stream might be large. A recursive function could overflow the stack in that case unless the recursive function were tail recursive and could be optimized to remove the recursive call.

A while loop in SML is written as while *Expr* do *Expr*. As is usual with while loops, the first *Expr* must evaluate to a boolean value. If it evaluates to true then the second *Expr* is evaluated. This process is repeated until the first *Expr* returns false.

5.20 Exception Handling

An exception occurs in SML when a condition occurs that requires special handling. If no special handling is defined for the condition the program terminates. As with most modern languages, SML has facilities for handling these exceptions and for raising user-defined exceptions.

Example 5.42

Consider the maxIntList function you wrote in practice problem 5.13. You probably had to figure out what to do if an empty list was passed to the function. One
way to handle this is to raise an exception.

```
1   exception emptyList;
2
3   fun maxIntList [] = raise emptyList
4     | maxIntList (h::t) = Int.max(h,maxIntList t) handle
5                                  emptyList => h
```

Invoking the maxIntList on an empty list can be handled using an exception
handling expression. The handle clause uses pattern matching to match the right
exception handler. To handle any exception the pattern _ can be used. The underscore
matches anything. Multiple exceptions can be handled by using the vertical bar (i.e.
|) between the handlers.

5.21 Encapsulation in ML

ML provides two language constructs that enable programmers to define new
datatypes and hide their implementation details. The first of these language constructs we'll look at is the signature. The other construct is the structure.

Signatures

A signature is a means for specifying a set of related functions and types without
providing any implementation details. This is analogous to an interface in Java or a
template in C++. Consider the datatype consisting of a set of elements. A set is a
group of elements with no duplicate values. Sets are very important in many areas of
Computer Science and Mathematics. Set theory is an entire branch of mathematics.
If we wanted to define a set in ML we could write a signature for it as follows.

Example 5.43

This is the signature of a group of set functions and a set datatype. Notice this
datatype is parameterized by a type variable so this could be a signature for a set
of anything. You'll also notice that while the type parameter is $'a$ there are type
variables named $''a$ within the signature. This is because some of these functions
rely on the equals operator. In ML the equals operator is polymorphic and cannot
be instantiated to a type. When this signature is used in practice the $'a$ and $''a$
types will be correctly instantiated to the same type.

```
1  signature SetSig =
2  sig
3      exception Choiceset
4      exception Restset
5      datatype 'a set = Set of 'a list
6      val emptyset    : 'a set
7      val singleton   : 'a -> 'a set
8      val member      : ''a -> ''a set -> bool
9      val union       : ''a set -> ''a set -> ''a set
10     val intersect    : ''a set -> ''a set -> ''a set
11     val setdif      : ''a set -> ''a set -> ''a set
12     val card        : 'a set -> int
13     val subset      : ''a set -> ''a set -> bool
14     val simetdif     : ''a set -> ''a set -> ''a set
15     val forall      : ''a set -> (''a -> bool) -> bool
16     val forsome      : ''a set -> (''a -> bool) -> bool
17     val forsomeone   : 'a set -> ('a -> bool) -> bool
18 end
```

Before a signature can be used, each of these functions must be implemented in a structure that implements the signature. This encapsulation allows a programmer to write code that uses these set functions without regards to their implementation. An implementation must be provided before the program can be run. However, if a better implementation comes along later it can be substituted without changing any of the code that uses the set signature.

Implementing a Signature

To implement a signature we can use the struct construct that we've seen before. In this case it is done as follows.

Example 5.44

Here is an implementation of the set signature.

```
1  (***** An Implementation of Sets as a SML datatype *****)
2
3  structure Set : SetSig =
4  struct
5
6  exception Choiceset
7  exception Restset
8
9  datatype 'a set = Set of 'a list
10
11 val emptyset = Set []
12
13 fun singleton e = Set [e]
14
```

```
15   fun member e (Set [])      = false
16     | member e (Set (h::t)) = (e = h) orelse member e (Set t)
17
18   fun notmember element st = not (member element st)
19
20   fun union (s1 as Set L1) (s2 as Set L2) =
21       let fun noDup e = notmember e s2
22       in
23           Set ((List.filter noDup L1)@(L2))
24       end
25
26   ...
27   end
```

Of course, the entire implementation of all the set functions in the signature is required. Some of these functions are left as an exercise.

☞ Practice 5.24

1. Write the card function. Cardinality of a set is the size of the set.
2. Write the intersect function. Intersection of two sets are just those elements that the two sets have in common. Sets do not contain duplicate elements.

5.22 Type Inference

Perhaps Standard ML's strongest point is the formally proven soundness of its type inference system. ML's type inference system is guaranteed to prevent any run-time type errors from occurring in a program. This turns out to prevent many run-time errors from occurring in your programs. Projects like the Fox Project[14] have shown that ML can be used to produce highly reliable large software systems.

The origins of type inference include Haskell Curry and Robert Feys who in 1958 devised a type inference algorithm for the simply typed lambda calculus. In 1969 Roger Hindley worked on extending this type inference algorithm. In 1978 Robin Milner independently from Hindley devised a similar type inference system proving its soundness. In 1985 Luis Damas proved Milner's algorithm was complete and extended it to support polymorphic references. This algorithm is called the Hindley-Milner type inference algorithm or the Milner-Damas algorithm. The type inference system is based on a very powerful concept called unification.

Unification is the process of using type inference rules to bind type variables to values. The type inference rules look something like this.

(IF-THEN)

$$\frac{\varepsilon \vdash e_1 : bool, \ \varepsilon \vdash e_2 : \alpha, \ \varepsilon \vdash e_3 : \beta, \ \alpha = \beta}{\varepsilon \vdash if \ e_1 \ then \ e_2 \ else \ e_3 : \alpha}$$

This rule says that for an if-then expression to be correctly typed, the type of the first expression must be a `bool` and the types of the second and third expression must be unifiable. If those preconditions hold, then the type of the if-then expression is given by the type of either of the second two expressions (since they are the same). Unification happens when $\alpha = \beta$ in the rule above. The ε is the presence of type information that is used when determining the types of the three expressions and is called the type environment.

Here are two examples that suggest how the type inference mechanism works.

Example 5.45

In this example we determine the type of the following function.

```
fun f(nil,nil) = nil
  | f(x::xs,y::ys) = (x,y)::f(xs,ys);
```

The function f takes one parameter, a pair.

```
f: 'a * 'b -> 'c
```

From the nature of the argument patterns, we conclude that the three unknown types must be lists.

```
f: ('p list) * ('s list) -> 't list
```

The function imposes no constraints on the domain lists, but the codomain list must be a list of pairs because of the cons operation $(x, y) ::$. We know $x:'p$ and $y:'s$. Therefore $'t='p*'s$.

```
f: 'p list * 's list -> ('p * 's) list
```

where $'p$ and $'s$ are any ML types.

Example 5.46

In this example the type of the function g is inferred.

```
1  fun g h x = if null x then nil
2              else
3                 if h (hd x) then g h (tl x)
4                 else (hd x)::g h (tl x);
```

The function g takes two parameters, one at a time.

```
g: 'a -> 'b -> 'c
```

The second parameter, x, must serve as an argument to `null`, `hd`, and `tl`; it must be a list.

```
g: 'a -> ('s list) -> 'c
```

The first parameter, h, must be a function since it is applied to `hd x`, and its domain type must agree with the type of elements in the list. In addition, h must produce a boolean result because of its use in the conditional expression.

```
g: ('s -> bool) -> ('s list) -> 'c
```

The result of the function must be a list since the base case returns `nil`. The result list is constructed by the code `(hd x)::g h (tl x)`, which adds items of type ′s to the resulting list.

Therefore, the type of g must be:

```
g: ('s -> bool) -> 's list -> s list
```

5.23 Exercises

In the exercises below you are encouraged to write other functions that may help you in your solutions. You might have better luck with some of the harder ones if you solve a simpler problem first that can be used in the solution to the harder problem.

You may wish to put your solutions to these problems in a file and then

```
- use "thefile";
```

in SML. This will make writing the solutions easier. You can try the solutions out by placing tests right within the same file. You should always comment any code you write. Comments in SML are preceded with a (* and terminated with a *) .

1. Reduce $(\lambda z.z + z)((\lambda x.\lambda y.x + y)\ 4\ 3)$ by normal order and applicative order reduction strategies. Show the steps.
2. How does the SML interpreter respond to evaluating each of the following expressions? Evaluate each of these expression in ML and record what the response of the ML interpreter is.

   a. 8 div 3;
   b. 8 mod 3;
   c. "hi"^"there";
   d. 8 mod 3 = 8 div 3 orelse 4 div 0 = 4;
   e. 8 mod 3 = 8 div 3 andalso 4 div 0 = 4;

3. Describe the behavior of the orelse operator in exercise 2 by writing an equivalent if then expression. You may use nested if expressions. Be sure to try your solution to see you get the same result.
4. Describe the behavior of the andalso operator in exercise 2 by writing an equivalent if then expression. Again you can use nested if expressions.
5. Write an expression that converts a character to a string.
6. Write an expression that converts a real number to the next lower integer.
7. Write an expression that converts a character to an integer.
8. Write an expression that converts an integer to a character.
9. What is the signature of the following functions? Give the signature and an example of using each function.

   a. hd
   b. tl
   c. explode
   d. concat
   e. :: - This is an infix operator. Use the prefix form of op :: to get the signature.

10. The greatest common divisor of two numbers, x and y, can be defined recursively. If y is zero then x is the greatest common divisor. Otherwise, the greatest common divisor of x and y is equal to the greatest common divisor of y and the remainder x divided by y. Write a recursive function called gcd to determine the greatest common divisor of x and y.

11. Write a recursive function called `allCaps` that given a string returns a capitalized version of the string.

12. Write a recursive function called `firstCaps` that given a list of strings, returns a list where the first letter of each of the original strings is capitalized.

13. Using pattern matching, write a recursive function called `swap` that swaps every pair of elements in a list. So, if `[1,2,3,4,5]` is given to the function it returns `[2,1,4,3,5]`.

14. Using pattern matching, write a function called `rotate` that rotates a list by n elements. So, `rotate(3,[1,2,3,4,5])` would return `[4,5,1,2,3]`.

15. Use pattern matching to write a recursive function called `delete` that deletes the nth letter from a string. So, `delete(3,"Hi there")` returns `"Hi here"`. HINT: This might be easier to do if it were a list.

16. Again, using pattern matching write a recursive function called `power` that computes x^n. It should do so with $O(log\ n)$ complexity.

17. Rewrite the `rotate` function of exercise 14 calling it `rotate2` to use a helper function so as to guarantee $O(n)$ complexity where n is the number of positions to rotate.

18. Rewrite exercise 14's `rotate(n,lst)` function calling it `rotate3` to guarantee that less than l rotations are done where l is the length of the list. However, the outcome of rotate should be the same as if you rotated n times. For instance, calling the function as `rotate3(6,[1,2,3,4,5])` should return `[2,3,4,5,1]` with less than 5 recursive calls to `rotate3`.

19. Rewrite the `delete` function from exercise 15 calling it `delete2` so that it is curried.

20. Write a function called `delete5` that always deletes the fifth character of a string.

21. Use a higher-order function to find all those elements of a list of integers that are even.

22. Use a higher-order function to find all those strings that begin with a lower case letter.

23. Use a higher-order function to write the function `allCaps` from exercise 11.

24. Write a function called `find(s,file)` that prints the lines from the file named `file` that contain the string s. You can print the lines to `TextIO.stdOut`. The `file` should exist and should be in the current directory.

25. Write a higher-order function called `transform` that applies the same function to all elements of a list transforming it to the new values. However, if an exception occurs when transforming an element of the list, the original value in the given list should be used. For instance,

```
- transform (fn x => 15 div x) [1,3,0,5]
val it = [15,5,0,3] : int list
```

26. The natural numbers can be defined as the set of terms constructed from 0 and the $succ(n)$ where n is a natural number. Write a datatype called `Natural` that can be used to construct natural numbers like this. Use the capital letter O for your zero value so as not to be confused with the integer 0 in SML.

27. Write a `convert(x)` function that given a natural number like that defined in exercise 26 returns the integer equivalent of that value.
28. Define a function called `add(x, y)` that given x and y, two natural numbers as described in exercise 26, returns a natural number that represents the sum of x and y. For example,

```
- add(succ(succ(0)),succ(0))
val it = succ(succ(succ(0))) : Natural
```

You may NOT use `convert` or any form of it in your solution.
29. Define a function called `mul(x, y)` that given x and y, two natural numbers as described in exercise 26, returns a natural that represents the product of x and y. You may NOT use `convert` or any form of it in your solution.
30. Using the `add` function in exercise 28, write a new function `hadd` that uses the higher order function called `foldr` to add together a list of natural numbers.

5.24 Solutions to Practice Problems

These are solutions to the practice problems . You should only consult these answers after you have tried each of them for yourself first. Practice problems are meant to help reinforce the material you have just read so make use of them.

Solution to Practice Problem 5.1

Addition is not commutative in Pascal or Java. The problem is that a function call, which may be one or both of the operands to the addition operator, could have a side-effect. In that case, the functions must be called in order. If no order is specified within expression evaluation then you can't even reliably write code with side-effects within an expression.

Here's another example of the problem with side-effects within code. In the code below, it was observed that when the code was compiled with one C++ compiler it printed 1,2 while with another compiler it printed 1,1. In this case, the language definition is the problem. The C++ language definition doesn't say what should happen in this case. The decision is left to the compiler writer.

```
int x = 1;
cout << x++ << x << endl;
```

The practice problem writes 17 as written. If the expression were b+a() then 15 would be written.

Solution to Practice Problem 5.2

With either normal order or applicative order function application is still left-associative. There is no choice for the initial redex.

$$(\lambda xyz.xz(yz))(\lambda x.x)(\lambda xy.x)$$
$$\Rightarrow (\lambda yz.(\lambda x.x)z(yz))(\lambda xy.x)$$
$$\Rightarrow (\lambda yz.z(yz))(\lambda xy.x)$$
$$\Rightarrow \lambda z.z((\lambda xy.x)z)$$
$$\Rightarrow \lambda z.z(\lambda y.z)\square$$

Solution to Practice Problem 5.3

Normal Order Reduction
$$(\lambda x.y)((\lambda x.xx)(\lambda x.xx))$$
$$\Rightarrow y$$
Applicative Order Reduction
$$(\lambda x.y)((\lambda x.xx)(\lambda x.xx))$$
$$\Rightarrow (\lambda x.y)((\lambda x.xx)(\lambda x.xx))$$

$$\Rightarrow (\lambda x.y)((\lambda x.xx)(\lambda x.xx))$$
$$\Rightarrow (\lambda x.y)(\overline{(\lambda x.xx)(\lambda x.xx)})$$
...

You get the idea.

Solution to Practice Problem 5.4

```
x div 6
Real.round(Real.fromInt(x) * y)
x / 6.3
x mod y
```

Solution to Practice Problem 5.5

```
fun factorial(n) = if n=0 then 1 else n*factorial(n-1)
```

Solution to Practice Problem 5.6

The recursive definition is $fib(0) = 0$, $fib(1) = 1$, $fib(n) = fib(n-1) + fib(n-2)$. The recursive function is:

```
fun fib(n) = if n = 0 then 1 else
             if n = 1 then 1 else
             fib(n-1) + fib(n-2)
```

Solution to Practice Problem 5.7

The solutions below are example solutions only. Others exist. However, the problem with each invalid list is not debatable.

1. You cannot cons a character onto a string list.
   `"a"::["beautiful day"]`
2. You cannot cons two strings. The second operand must be a list.
   `"hi"::["there"]`
3. The element comes first in a cons operation and the list second.
   `"you"::["how","are"]`
4. Lists are homogeneous. Reals and integers can't be in a list together.
   `[1.0,2.0,3.5,4.2]`
5. Append is between two lists.
   `2::[3,4]` or `[2]@[3,4]`

6. Cons works with an element and a list, not a list and an element.

```
3::[]
```

Solution to Practice Problem 5.8

```
fun explode(s) =
  if s = "" then []
  else String.sub(s,0)::
       (explode(String.substring(s,1,String.size(s)-1)))
```

Solution to Practice Problem 5.9

```
fun reverse(L) =
  if null L then []
  else append(reverse(tl(L)),[hd(L)])
```

Solution to Practice Problem 5.10

```
fun reverse([]) = []
  | reverse(h::t) = reverse(t)@[h]
```

Solution to Practice Problem 5.11

```
1  let val x = 10
2  in
3      (* 1. Value of x = 10 *)
4      let val x = x+1
5      in
6          (* 2. Value of x = 11 (hidden x still is 10) *)
7          x
8      end;
9      (* 3. Value of x = 10 (hidden x is visible again) *)
10     x
11 end
```

Solution to Practice Problem 5.12

```
datatype intlist = nil' | cons of int * intlist;
```

Solution to Practice Problem 5.13

```
fun maxIntList nil' = valOf(Int.minInt)
  | maxIntList (cons(x,xs)) = Int.max(x,maxIntList xs)
```

or

```
fun maxIntList (cons(x,nil')) = x
  | maxIntList (cons(x,xs)) = Int.max(x,maxIntList xs)
```

The second solution will cause a pattern match nonexhaustive warning. That should be avoided, but is OK in this case. The second solution will raise a pattern match exception if an empty list is given to the function. See the section on exception handling for a better solution to this problem.

Solution to Practice Problem 5.14

The first step in the solution is to determine the number of calls required for values of n. Consulting figure 5.1 shows us that the number of calls are 1, 1, 3, 5, 9, 15, 25, etc. The next number in the sequence can be found by adding together two previous plus one more for the initial call.
The solution is that for $n \geq 3$ the function 1.5^n bounds the number of calls on the lower side while 2^n bounds it on the upper side. Therefore, the number of calls increases exponentially.

Solution to Practice Problem 5.15

The cons operation is called n times where n is the length of the first list when append is called. When reverse is called it calls append with $n-1$ elements in the first list the first time. The first recursive call to reverse calls append with $n-2$ elements in the first list. The second recursive call to reverse calls append with $n-3$ elements in the first list. If we add up $n-1+n-2+n-3+...$ we end up with $\sum_{i=1}^{n-1} i = ((n-1)n)/2$. Multiplying this out leads to an n^2 term and the overall complexity of reverse is $O(n^2)$.

Solution to Practice Problem 5.16

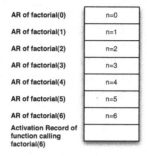

AR of factorial(0)	n=0
AR of factorial(1)	n=1
AR of factorial(2)	n=2
AR of factorial(3)	n=3
AR of factorial(4)	n=4
AR of factorial(5)	n=5
AR of factorial(6)	n=6
Activation Record of function calling factorial(6)	

Fig. 5.2: The run-time stack when factorial(6) is called at its deepest point

Solution to Practice Problem 5.17

This solution uses the accumulator pattern and a helper function to implement a linear time reverse.

```
1  fun reverse(L) =
2      let fun helprev (nil, acc) = acc
3            | helprev (h::t, acc) = helprev(t,h::acc)
4      in
5        helprev(L,[])
6      end
```

Solution to Practice Problem 5.18

This solution is surprisingly hard to figure out. In the first, f is certainly an uncurried function (look at how it is applied). The second requires f to be curried.

```
- fun curry f x y = f(x,y)
val curry = fn : ('a * 'b -> 'c) -> 'a -> 'b -> 'c

- fun uncurry f (x,y) = f x y
val uncurry = fn : ('a -> 'b -> 'c) -> 'a * 'b -> 'c
```

Solution to Practice Problem 5.19

The first takes a list of lists of integers and adds one to each integer of each list in the list of lists.

The second function takes a list of functions that all take the same type argument, say a'. The function returns a list of functions that all take an a' list argument. The example below might help. The list of functions that is returned by (map map) is suitable to be used as an argument to the construction function discussed earlier in the chapter.

```
- map (map add1);
val it = fn : int list list -> int list list

(map map);
stdIn:63.16-64.10 Warning: type vars not generalized because
of value restriction are instantiated to dummy types
(X1,X2,...)
val it = fn : (?.X1 -> ?.X2) list ->
                    (?.X1 list -> ?.X2 list) list
- fun double x = 2 * x;
val double = fn : int -> int
- val flist = (map map) [add1,double];
val flist = [fn,fn] : (int list -> int list) list
- construction flist [1,2,3];
val it = [[2,3,4],[2,4,6]] : int list list
```

Solution to Practice Problem 5.20

foldl is left-associative and foldr is right-associative.

```
- foldr op :: nil [1,2,3];
val it = [1,2,3] : int list
- foldr op @ nil [[1],[2,3],[4,5]];
val it = [1,2,3,4,5] : int list
```

Solution to Practice Problem 5.21

```
- List.filter (fn x => x mod 7 = 0) [2,3,7,14,21,25,28];
val it = [7,14,21,28] : int list
- List.filter (fn x => x > 10 orelse x = 0)
            [10, 11, 0, 5, 16, 8];
val it = [11,0,16] : int list
```

Solution to Practice Problem 5.22

```
cpslen [1,2,3] (fn v => v)
= cpslen [2,3] (fn w => ((fn v => v) (1 + w)))
= cpslen [3]
        (fn x => ((fn w => ((fn v => v) (1 + w)))(1 + x)))
= cpslen []
        (fn y => ((fn x => ((fn w => ((fn v => v)
        (1 + w)))(1 + x)))(1 + y)))
= (fn y => ((fn x => ((fn w => ((fn v => v)
        (1 + w)))(1 + x)))(1 + y))) 0
= (fn x => ((fn w => ((fn v => v) (1 + w)))(1 + x))) 1
= (fn w => ((fn v => v) (1 + w))) 2
= (fn v => v) 3
= 3
```

Solution to Practice Problem 5.23

```
1   datatype bintree = termnode of int
2            | binnode of int * bintree * bintree;
3
4   val tree = (binnode(5,binnode(3,termnode(4),binnode(8,
5                termnode(5),termnode(4))), termnode(4)));
6
7   fun depth (termnode _) = 0
8     | depth (binnode(_,t1,t2)) = Int.max(depth(t1),depth(t2))+1
9
10  fun cpsdepth (termnode _) k = k 0
11    | cpsdepth (binnode(_,t1,t2)) k =
12        Int.max(cpsdepth t1 (fn v => (k (1 + v))),
13                cpsdepth t2 (fn v => (k (1 + v))))
```

Solution to Practice Problem 5.24

```
1   fun card (Set L) = List.length L;
2
3   fun intersect (Set L1) S2 =
4        Set ((List.filter (fn x => member x S2) L1))
```

5.25 Additional Reading

Jeffrey Ullman's book on functional programming in Standard ML [36] is a very good introduction and reference for Standard ML. It is more thorough than the topics provided in this text and contains many topics not covered here including discussion of arrays, functors, and sharings along with a few of the Basis structures. The topics presented here and in the next chapter give you a good introduction to the ideas and concepts associated with functional programming. Given the online resources available for the Standard ML Basis library, Jeffrey Ullman's book, and the information present here, there should be enough for you to become a very proficient ML programmer.

Chapter 6

Language Implementation in Standard ML

The ML in the name *Standard ML* stands for meta-language. SML was designed as a language for describing languages when it was used as part of the Logic for Computable Functions (LCF) system. That tradition remains in place today. Standard ML is an excellent language choice for implementing interpreters and compilers. There are two very nice tools for SML that will generate a scanner and a parser given a description of the tokens and grammar of a language. The scanner generator is called ML-lex and the parser generator is called ML-yacc. This chapter introduces these two tools through a case study involving the development of a simple compiler for the calculator language presented in previous chapters.

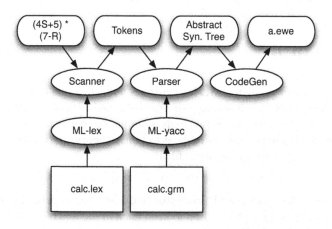

Fig. 6.1: Structure of the Calculator Compiler implemented in ML

Figure 6.1 depicts all the relevant pieces in constructing and using the expression compiler. The compiler begins by calling a function which calls the parser. The

K.D. Lee, *Programming Languages*, DOI: 10.1007/978-0-387-79421-1_6,
© Springer Science+Business Media, LLC 2008

parser returns an abstract syntax tree of the expression being evaluated. The parser gets tokens from the scanner to parse the input and build the AST. The scanner reads characters from the input file and groups them together as tokens. When an AST is returned by the parser, the compiler calls the code generator to evaluate the tree and produce the target code. The target code for this project will be EWE code. In this chapter the scanner and the parser won't have to be written by hand. ML-lex and ML-yacc will be used to generate these parts of the compiler from specifications that are provided to these tools.

There are two main parts to a compiler, the front end and back end. The front end reads the tokens and builds an AST of a program. The back end generates the code given the AST representation of the program. As presented in earlier chapters, the front end consists of the scanner and the parser. ML-lex is a tool that given a definition of the tokens of a language will generate a scanner. ML-yacc is a tool that given a grammar will generate a parser. This chapter discusses a partial implementation of the front end and back end for a calculator language compiler. In this case study, ML-lex and ML-yacc use the following files to generate the front end.

1. The tokens of the language are defined in a file called `calc.lex`.
2. The datatype for expression ASTs is defined in a file called `calcast.sml`.
3. The grammar of the language is defined in a file called `calc.grm`. This file also contains a mapping from productions in the grammar to nodes in an AST. The parser reads tokens and builds an AST of the expression being compiled.

The next sections will introduce ML-lex and ML-yacc. By the end of the chapter you will have enough information to complete the compiler using SML. Later you will be given the opportunity to extend the calculator language and compiler.

6.1 Using ML-lex

ML-lex is a scanner generator. ML-lex generates a function that can be used to get tokens from the input. It is based on a similar tool called *lex* that generates scanners for C programs. The input to the two tools is similar but not exactly the same. The input to ML-lex is a file consisting of three sections, where each section is separated by `%%`. The format of an ML-lex input file is:

```
User declarations %% ML-lex definitions %% Rules
```

The user declarations are any ML code that will assist you in defining the tokens. Typically, a reference variable is used to keep track of the line of input being read. There might also be some functions for converting strings to other values like integers. An error function that handles bad tokens is a common function for this section to get the scanner and the parser to work together.

The ML-lex declarations follow the user declarations. Sets of characters are declared in this section. In addition a functor must be declared. A functor is a module that takes a structure as a parameter and returns a new structure as a result. A functor is used by ML-lex and ML-yacc to create the scanner.

The last section of an ML-lex definition is composed of a set of rules that define the tokens of the language. Each rule has the form:

```
{reg_exp} => (return_value);
```

The reg_exp is a regular expression. The language of regular expressions can be used to define tokens. Every regular expression can be expressed as a finite state machine. Finite state machines can be used to recognize tokens. The set of reg_exp is eventually translated into a finite state machine that can be used to recognize tokens in the language. When a string of characters is recognized as a token, its matching return value is constructed from the rules and that value is returned by the scanner to the parser. Seeing an example will help make some sense of this.

Example 6.1

Here is an ML-lex specification for the calculator language.

```
1  type pos = int
2  type svalue = Tokens.svalue
3  type ('a, 'b) token = ('a, 'b) Tokens.token
4  type lexresult = (svalue, pos) token
5  val pos = ref 1
6  val error = fn x => TextIO.output(TextIO.stdErr, x ^ "\n")
7  val eof = fn () => Tokens.EOF(!pos, !pos)
8  fun sval([], r) = r
9    | sval(a::s, r) = sval (s, r*10+(ord(a) - ord(#"0")));
10 %%
11 %header (functor calcLexFun(structure Tokens : calc_TOKENS));
12 alpha=[A-Za-z];
13 alphanumeric=[A-Za-z0-9_];
14 digit=[0-9];
15 ws=[\ \t];
16 %%
17 \n   => (pos := (!pos) + 1; lex());
18 {ws}+  => (lex());
19 "("   => (Tokens.LParen(!pos,!pos));
20 ")"   => (Tokens.RParen(!pos,!pos));
21 "+"   => (Tokens.Plus(!pos,!pos));
22 "*"   => (Tokens.Times(!pos,!pos));
23 "/"   => (Tokens.Div(!pos,!pos));
24 "-"   => (Tokens.Minus(!pos,!pos));
25 {digit}+  => (Tokens.Int(sval(explode yytext,0),!pos,!pos));
26 {alpha}{alphanumeric}* =>
27    (let val tok = String.implode (List.map (Char.toLower)
28              (String.explode yytext))
29     in
30        if     tok="s" then Tokens.Store(!pos,!pos)
31        else if tok="r" then Tokens.Recall(!pos,!pos)
32        else (error ("error: bad token "^yytext); lex())
33     end);
34 .   => (error ("error: bad token "^yytext); lex());
```

The svalue type must be defined in the user declarations. The token type also has to be defined. The `lexresult` variable defines tokens as having both an svalue and an integer position. These three declarations are required to define the signature of the scanner which is called the lexer by ML-lex.

In the ML-lex declarations the functor is declared as required by the parser. The alpha declaration declares a class of characters called `alpha` that consists of letters a to z in lower and upper case. The `alphanumeric` characters include letters and digits and underscores. The `digit` declaration defines the class of digits as being 0 to 9. The `ws` stands for whitespace. It defines blanks and tabs as whitespace.

Finally, the rules define all the tokens. The first two rules discard whitespace in the input. The \n matches the newline character and adds one to the line position when it is found. Instead of returning a value, if a newline is found the scanner calls lex (the scanner's getToken function) recursively to get another token. Single character tokens are defined in double quotes. Character classes may be used to define tokens. When a character class is used it is placed inside braces (i.e. `digit`). The + sign means one or more of the preceding class of characters. the * means zero or more of the preceding class of characters. So a keyword is an `alpha` character followed by zero or more `alphanumeric` characters.

In the last line the period matches any character, so anything that gets through to that rule must be a bad character. When *yytext* is referenced in the code that is the actual lexeme of the token.

From a definition like the calc.lex file shown here the ML-lex tool has enough information to generate a scanner for the tokens of the language. The parser will get tokens and build an AST for the expression. The SML definition of calculator ASTs is given in the next section.

☞ Practice 6.1

Given the ML-lex specification in example 6.1, what more would have to be added to allow expressions like this to be correctly tokenized by the scanner? What new tokens would have to be recognized? How would you modify the specification to accept these tokens?

```
1   let val x = 6
2   in
3       x + 6
4   end
```

6.2 The Calculator Abstract Syntax Definition

The parser reads tokens and builds an abstract syntax tree of a calculator expression. In SML, the abstract syntax definition is given by an SML datatype. Each type of node in the tree is tagged with its type. Some nodes in the tree include the subtrees such as the addition, subtraction, multiplication, division, and store nodes. The negation node is added to support unary negation in expressions. Now -5 is a valid expression. The recall and integer nodes are leaf nodes in any AST. They have no subtrees. The integer node keeps track of its integer value.

Example 6.2

This is the abstract syntax definition for calculator ASTs.

```
1  structure calcAS =
2  struct
3
4  datatype
5      AST = add' of AST * AST
6          | sub' of AST * AST
7          | prod' of AST * AST
8          | div' of AST * AST
9          | negate' of AST
10         | integer' of int
11         | store' of AST
12         | recall';
13 end;
```

☞ **Practice 6.2**

How would you modify the abstract syntax so expressions like the one below could be represented?

```
1  let val x = 6
2  in
3      x + 6
4  end
```

6.3 Using ML-yacc

ML-yacc is a parser generator. The name stands for *Yet Another Compiler Compiler* (i.e. yacc). *yacc* is a tool that generates parsers for C programs to use. ML-yacc is the SML version of this tool. ML-yacc is a little different than yacc but provides mostly the same functionality. ML-yacc's input format is similar to ML-lex's input format. An ML-yacc specification consists of three parts.

```
User declarations %% ML-yacc definitions %% Rules
```

Another example will assist in understanding the format of an ML-yacc specification.

Example 6.3

This is the ML-yacc specification for the calculator language. The file is called `calc.grm`.

```
1   open calcAS;
2
3   %%
4   %name calc (* calc becomes a prefix in functions *)
5   %verbose
6   %eop EOF
7   %pos int
8   %nodefault
9   %pure (* no side-effects in actions *)
10  %term EOF
11       | LParen
12       | RParen
13       | Plus
14       | Minus
15       | Times
16       | Div
17       | Store
18       | Recall
19       | Int of int
20  %nonterm Prog of Expr
21         | Expr of Expr
22         | Term of Expr
23         | StoreIt of Expr
24         | NegFactor of Expr
25         | Factor of Expr
26
27  %%
28  Prog : Expr                         (Expr)
29
30  Expr : Expr Plus Term               (add'(Expr,Term))
31       | Expr Minus Term              (sub'(Expr,Term))
32       | Term                         (Term)
33
34  Term : Term Times StoreIt           (prod'(Term,StoreIt))
35       | Term Div StoreIt             (div'(Term,StoreIt))
36       | StoreIt                      (StoreIt)
37
38  StoreIt : NegFactor Store           (store'(NegFactor))
39          | NegFactor                 (NegFactor)
40
41  NegFactor : Minus NegFactor         (negate'(NegFactor))
42            | Factor                  (Factor)
43
44  Factor : Int                        (integer'(Int))
45         | LParen Expr RParen         (Expr)
46         | Recall                     (recall')
```

The user declarations consist of one line. The `open calcAS` opens the structure called `calcAS` so the parser can access the AST definition without having to precede the name with the structure name each time. Without the open statement, each time an AST node was referred to the fully qualified name would be required. For instance calcAS.add′ would have to be written instead of writing add′ each time an add AST node was referred to in the code.

The ML-yacc declarations include a name to prefix functions in the scanner with, in this case calc. The verbose helps in debugging. eop says that EOF is the last token returned. This helps in terminating the parser. eop stands for end of parse. The pos type is redeclared here for use with the scanner.

The nodefault tells the parser not to insert tokens it thinks might have been left out. This helps in finding syntax errors earlier than they would be otherwise. If this were omitted the parser would insert a token when it is reasonably sure the program being parsed is missing a token.

The pure declarations says that the parser has no side-effects. It simply builds a tree and returns it. This means that ML-yacc can undo certain parsing operations if it needs to without fear of a side-effect not being undone. Finally, and most importantly the terminals and nonterminals of the language are declared.

In the rules section of the ML-yacc specification the productions are declared on the left. To the right of each production is a return value that is returned when that production is used in parsing. ML-yacc generates a bottom-up parser, so productions are used in the reverse order they would be used in a right-most derivation.

When you see a production like this:

```
Expr : Expr Plus Term                    (add'(Expr,Term))
```

it means when the production for addition is used an AST with add′ at the root is returned where the left and right subtrees are the ASTs that were returned from parsing the two subexpressions.

Example 6.3 is typical of an ML-yacc parser definition. The next section shows you how to use the scanner and the parser to build an AST and generate code for it.

☞ Practice 6.3

What modifications would be required in the calc.grm specification to parse expressions like the one below?

```
1   let val x = 6
2   in
3       x + 6
4   end
```

6.4 Code Generation

Code generation is essential to any compiler. The code generator translates the abstract syntax tree into a language that may either be executed directly or interpreted by some low-level interpreter like the Java Virtual Machine (i.e. JVM). In this case study, the code generator generates code for a register machine that is emulated using the EWE interpreter.

The code generator's code for addition and subtraction of integers is given in Appendix D. Lines 6-35 are required to create the parser and the scanner and tie them together. The `compile` function at the bottom of the code calls the parser and generates a little of the target program's code, which will be explained shortly. The `compile` function creates a file called a.ewe while will contain the compiled program.

The `run` function is required so the SML program (i.e. the compiler) can itself be compiled. The `run` function is exported from SML so it can be called from the command-line. The `run` function invokes the `compile` function passing it the name of the file containing the program to be compiled.

Most of the work in the compiler is performed by the `codegen` function. This function is responsible for generating EWE code for every possible calculator expression. This is accomplished by a postfix traversal of an expresssion's AST. In a postfix traversal of an AST the code is first generated for the left subtree (if there is one), then code is generated for the right subtree (again, if there is one). Finally, code is generated for the root node of the AST. This is a recursive definition so we can start by considering a very simple case. In fact, consider just the simplest case, an expression containing just one number, say 5.

Example 6.4

As presented in this chapter, the calculator compiler will compile some simple expressions. For instance, 5 is an expression that will compile. Compiling a program containing 5 yields the following EWE code.

```
1  SP:=100
2  R0:=5
3  writeInt(R0)
4  halt
5
6  equ MEM M[12]
7  equ SP M[13]
8  equ R0 M[0]
```

The goal when this program runs is to print 5 to the screen. This EWE program does that.

Because `codegen` is necessarily a recursive function, the result of generating code for an AST must be left someplace where it can be found. It's not good enough to just store the value in some variable because `codegen` is recursive. If code is generated for a left and right subtree in a postfix traversal of the tree and the code leaves the two values in the same location then the first result will be left in the same place as the second result.

You've likely dealt with this postfix traversal problem before. The solution is to use a stack. You can push the result of executing the code in the left subtree on a stack. The value of the right subtree can also be pushed. After executing the code for both subtrees, the top two values on the stack will be the two values needed to complete the calculation of the expression. Another example will clear things up.

Example 6.5

Consider generating code for 5 + 4. If we were to blindly follow the example above the EWE code would look something like this:

```
 1  SP:=100
 2  R0:=5
 3  R0:=4
 4  R0:=R0+R0
 5  writeInt(R0)
 6  halt
 7
 8  equ MEM M[12]
 9  equ SP M[13]
10  equ R0 M[0]
```

Obviously this program would print 8 as a result, not the 9 that we want. The problem is that we can't leave the 5 and the 4 in the same place. That suggests we want something like this to be generated instead.

```
 1  SP:=100
 2  R0:=5
 3  R1:=4
 4  R0:=R0+R1
 5  writeInt(R0)
 6  halt
 7
 8  equ MEM M[12]
 9  equ SP M[13]
10  equ R0 M[0]
11  equ R1 M[1]
```

The code in the example above suggests that the stack used to generate code somehow exists in R0, R1, R2, R3, and so on. When we want to push something on the stack we can just store it in the next Rn. In fact it's a little more complicated than that, but not much.

Most machine languages (or assembly languages, if you prefer) are register based. Temporary values are stored in registers so they can be accessed again quickly. Registers are simply named memory locations that exist inside the CPU of the computer. The EWE language can be used to simulate a register machine. The registers in EWE will be called R0, R1, R2, etc.

However, a stack seems to be the structure we want when generating code. It would be nice if somehow registers could be made to look like a stack. They can. Register allocation is a complex topic that goes beyond what is covered in a programming languages course, but with a little structure, it isn't too hard to allocate registers so they resemble pushing and popping from a stack. Thankfully, there is a register allocation framework that does just that, it emulates a stack.

There are four functions defined by this register allocation framework. When a temporary location is needed to store a value, a register can be allocated using the `getReg` function. When a register needs to be pushed on the stack the `pushReg` function is used. To get a register off the register stack the `popReg` function is used. Finally, to delete a register that is no longer needed the `delReg` function is used.

Example 6.6

Here is the register allocation framework in action. This example, taken from Appendix D, shows how code is generated to add two values together.

```
1   fun codegen(add'(t1,t2),outFile,bindings,offset,depth) =
2       let val _ = codegen(t1,outFile,bindings,offset,depth)
3           val _ = codegen(t2,outFile,bindings,offset,depth)
4           val reg2 = popReg()
5           val reg1 = popReg()
6       in
7         TextIO.output(outFile,reg1^":="^reg1^"+"^reg2^"\n");
8         delReg(reg2);
9         pushReg(reg1)
10      end
11
12    | codegen(integer'(i),outFile,bindings,offset,depth) =
13       let val r = getReg()
14       in
15         TextIO.output(outFile, r^":="^Int.toString(i)^"\n");
16         pushReg(r)
17      end
```

Generating code for 5 + 4 first generates code for the 5. Above, the code generation for 5 gets a register using `getReg`, writes some code to the EWE program to put the 5 in the register, and then pushes the register on the stack using `pushReg`.

Next, code for 4 is generated in the same way resulting in R0 holding the 5 and R1 holding the 4. The register stack has R0 on the bottom with R1 on top of it. The `bindings`, `offset`, and `depth` parameters will be discussed later.

The code generation for adding the two values together first generates the code for the two subtrees. When that is completed, R0 and R1 are on the register stack with R1 on top. The `popReg` function is called twice to get those registers off the stack. EWE code is generated to add the two values together. R0 is used again to store the value since it is no longer needed to hold the 5. The R1 register is not needed anymore so it is deleted with a call to `delReg`. Finally, R0 is pushed on the stack since that value needs to be saved for later.

☞ Practice 6.4

How can code be generated to multiply two numbers together? How can code be generated to negate a value as in unary negation? Refer back to the AST definition to see what these nodes in an AST would look like. You can also refer back to the EWE language definition. Follow the pattern in example 6.6 to generate the correct code for expressions containing multiplication and unary negation.

There are a few things that should be pointed out about the register allocation framework. First, all registers that are allocated with getReg must eventually be freed with delReg. If this is not done, the framework will signal an error telling the programmer that an unfreed register still exists. In the compiler presented in this chapter, the very last register, containing the result of the expression, is freed by the compile function.

The register allocation framework insists that all registers are allocated and freed in a first in/first out fashion. This means that if R0 is allocated before R1 then R0 should be freed after R1. This isn't too restrictive since the allocation framework is mimicking a stack anyway.

Finally, it should be noted that this framework is a compile-time simulation of a run-time stack. When registers are allocated and freed, this occurs at compile time. The registers being allocated or freed are usually called symbolic registers. Symbolic registers are mapped to real registers (memory locations in EWE) by constructing a register interference graph and through a process called graph coloring. Thankfully, the register allocation framework presented here does this work for us.

6.5 Compiling in Standard ML

The SML Calculator compiler presented in this chapter can be compiled and run to add and subtract integers without any modifications. The rest of the compiler implementation is left as an exercise. A tool called the compiler manager (CM) is used to compile code in Standard ML. The compiler manager looks at the date of source files and compiled files to determine which ML source files need to be compiled, much like the Unix Make tool. A file called sources.cm tells CM the name of your source files. In this case study, the standard Unix Make tool is used to get everything started. To compile the project you enter the command make at the command-line.

> **Example 6.7**
>
> This is the Makefile for the compiler project. It invokes the Makefile.gen script to start the SML compilation process. The CM does the rest. The clean rule erases all files generated during compilation of the project.

```
1   all:
2           Makefile.gen;
3
4   clean:
5           rm calccomp*
6           rm calc.lex.sml
7           rm calc.grm.sml
8           rm calc.grm.desc
9           rm calc.grm.sig
10          rm -Rf CM
11          rm -Rf .cm
```

The Makefile says to invoke the Makefile.gen Unix script. A Unix script is a text file where the first line indicates an interpreter that should be used to *run* the program contained in the file. Unix provides this ability to run interpreters by invoking a text file containing the name of the interpreter to run.

Example 6.8

Makefile.gen is a Unix script. The C Shell interpreter called csh is used to run this file. The second line of the script says to start SML and read subsequent lines as if they were typed directly into the SML shell until the EOF token is encountered.

```
1  #!/bin/csh
2  sml << EOF
3  CM.make "sources.cm";
4  SMLofNJ.exportFn("calccomp",calc.run);
5  EOF
```

To invoke the SML compiler manager a function called CM.make is called with the name of a text file containing the names of the SML modules to compile. When invoked, the SMLofNJ.exportFn function tells SML to build a binary representation of the exported function and all code that the exported function is dependent on. The ability to export only the code needed by a program is unique among compilers. To know which modules to compile, the SML compiler manager must know which modules contain the code to compile.

Example 6.9

The sources.cm file is used by the SML compiler manager to know which modules to compile. The Group is line is always the first line. The $/basis.cm and the $/ml-yacc-lib.cm are required to tell the compiler manager that these two built-in libraries are to be included in the compilation. The rest of the lines are the modules of the compiler.

```
1  Group is
2    $/basis.cm
3    $/ml-yacc-lib.cm
4    calc.lex
5    calc.grm
6    calc.sml
7    calcast.sml
8    registers.sml
```

A second Unix script is used to start the compiler. The script is called calc. Typing calc at the command-line and pressing enter will prompt for a filename containing a calculator expression.

Example 6.10

This is the calculator compiler startup script, called `calc`. It reads the name of a text file from the user and then invokes SML by loading the binary image of the program created when the compiler was compiled by the compiler manager.

```
1  #!/bin/bash
2  set -f
3  echo -n "Enter an file name: "
4  read file
5  sml @SMLload=calccomp $file
```

6.6 Extending the Language

The calculator language can be extended in a variety of ways with very few changes. For instance, consider adding the ability to read input from the keyboard when the program runs. The addition of the keyword `get` could read an integer from the keyboard.

Example 6.11

To read a value from the keyboard while evaluating a calculator expression the lexical specification in `calc.lex` must be modified to add the `get` keyword to the list of keyword tokens. Keywords are added to the `alpha` followed by zero or more `alphanumeric` characters rule. That rule would be modified as shown here.

```
1  {alpha}{alphanumeric}* =>
2      (let val tok = String.implode (List.map (Char.toLower)
3                  (String.explode yytext))
4       in
5          if      tok="s" then Tokens.Store(!pos,!pos)
6          else if tok="r" then Tokens.Recall(!pos,!pos)
7          else if tok="get" then Tokens.Get(!pos,!pos)
8          else (error ("error: bad token "^yytext); lex())
9       end);
```

The `Get` token must be added to the terminal section of the `calc.grm` file and a production must be added to the rules in `calc.grm` to allow `get` to appear as a `Factor` in an expression.

```
Factor : Get                   (get')
```

The AST definition in `calcast.sml` must be modified so `get'` can appear as an AST node. Finally, the code generator in `calc.grm` must be modified to generate code for a `get'` node.

```
1  | codegen(get',outFile,bindings,offset,depth) =
2    let val r = getReg()
3    in
4      TextIO.output(outFile, "readInt(""^r^"")\n");
5      pushReg(r)
6    end
```

In this short example, the language was extended to allow input to be read from the keyboard. Implementing this makes it possible for a programmer to write a short program to read values from the keyboard and use them in an expression. This was accomplished by changing the scanner, parser, AST definition, and the code generator. Other more interesting extensions to the language are also possible.

6.7 Let Expressions

In chapter 5 let expressions were introduced as part of Standard ML. Let expressions enable a programmer to bind identifiers to values. The scope of an identifier is the body of its let expression.

Example 6.12

Consider the let expression given here. The scope of the identifier x is limited to the let expression's body.

```
1  let val x = 6
2  in
3    x + 5
4  end
```

The general format of a let expression is

```
let val id = {Expr}
in
   {Expr}
end
```

The goal is to generate code for both expressions in the *let* construct. The result of evaluating the first expression must be stored someplace where it can be referred to in the second expression.

As in the previous example the scanner, parser, AST definition, and code generator must be altered to support compilation of this type of expression. Code generation is the most difficult change. The change to the scanner is perhaps the second most difficult change. New tokens must be recognized by the scanner.

One of the new tokens is an identifier. An identifier is any string of characters that is not a keyword of the language. To recognize ID tokens the calc.lex file must be altered to return a Token.ID token when a string of characters is not a keyword rather than printing an error message.

Example 6.13

Currently the scanner specification in `calc.lex` checks for keywords and returns an error if a string of characters is not a keyword.

```
1  {alpha}{alphanumeric}* =>
2    (let val tok = String.implode (List.map (Char.toLower)
3                   (String.explode yytext))
4    in
5       if      tok="s" then Tokens.Store(!pos,!pos)
6       else if tok="r" then Tokens.Recall(!pos,!pos)
7       else (error ("error: bad token "^yytext); lex())
8    end);
```

The else clause needs to be modified so an identifier is returned instead of an error message. An appropriate replacement line for `calc.lex` would be

```
else Tokens.ID(yytext,!pos,!pos)
```

The `yytext` field is included with the `ID` token because the compiler will need to know what identifier was found. The `yytext` field in example 6.12 is `"x"`, the lexeme of the identifier. The other changes required to the scanner are straightforward and follow the patterns presented earlier.

There are other changes to `calc.lex`. These changes are outlined in the solution to practice problem 6.1 at the end of the chapter. The AST definition must change so let expressions can be represented in the abstract syntax.

Example 6.14

The additional AST node types required for let expressions are as follows. The `valref'` node occurs when a bound value is referred to in a program.

```
| letval' of string * AST * AST
| valref' of string
```

The parser changes are also similar to past examples and are given in the solution to practice problem 6.3 at the end of the chapter. A let expression is just another `Factor` of an expression in the grammar. Code generation is where things get interesting. Obviously, code needs to be generated for the two expressions that make up a let expression. The result of the first expression must be stored in a location that can be referred to when the identifier is used in the second expression. This constitutes the need for something called an activation record. An activation record is a piece of memory pointed to by a special register called the stack pointer (SP). Earlier examples in this chapter have included an SP register that is assigned to location 100 in memory. Now it is time to use the SP register.

The result of the first expression can be stored relative to the SP register. The compiler can keep track of the location of the value from the first expression and pass that along to the evaluation of the second expression. That is the purpose of a binding in a compiler. A binding is a pairing of an identifier and a location where the value represented by the identifier can be found.

Example 6.15

Consider the let expression given in example 6.12. The code that should be generated to evaluate that let expression is given here.

```
1   SP:=100
2   R0:=6
3   M[SP+0]:=R0
4   R1:=M[SP+0]
5   R2:=5
6   R1:=R1+R2
7   writeInt(R1)
8   halt
9
10  equ MEM M[12]
11  equ SP  M[13]
12  equ R0  M[0]
13  equ R1  M[0]
14  equ R2  M[1]
```

Line 1 initiallizes the stack pointer (SP). Line 2 loads the 6 into a register. Line 3 stores the 6 in the activation record at offset 0. This is done so the value x is bound to can be loaded back into a register and used. Line 4 is the beginning of the body of the let expression's code. It loads the 6 back into a register again. Line 7 loads the 5. Line 8 adds the two together.

Nowhere in example 6.15 does the identifier x appear. The identifier is not directly bound to the value in the EWE code. The identifier is bound in the compiler to a location where the value can be found. In this way bindings are eliminated from the target program. Bindings only exist in the compiler, not in the generated code. The compiler generates code that evaluates the identifier's expression and then stores that value in the activation record. This location in the activation record is bound to the identifier in the compiler. The binding is then used by the compiler when it comes time to generate code that refers to the bound value.

A binding is a triple of (identifier, offset, depth). The identifier is tagged with its type. In example 6.12 the binding contains constant' ("x") as the identifier portion. The constant' tag differentiates it from a function identifier which is allowed in other versions of the compiler. The offset portion of the binding is a string. In exampe 6.12 the offset portion is "0", the amount to add to the SP register to find the bound value in the activation record. The depth is not used in this version of the compiler and can just remain depth.

The bindings are a list of the binding values with the latest or newest bindings added to the front of the list. If a binding should be visible in a piece of code, the bindings passed when that code is generated should include the binding. The boundTo function looks up a binding in the list of bindings and returns its offset if found.

Example 6.16

Here is the code of the `boundTo` function.

```
1   exception unboundId;
2
3   datatype Type = function' of string
4                 | constant' of string;
5
6   fun boundTo(name,[]) =
7       let val idname = (case name of
8                           function'(s) => s
9                         | constant'(s) => s)
10      in
11         TextIO.output(TextIO.stdOut, "Unbound identifier "^
12                       idname^" referenced or type error!\n");
13         raise unboundId
14      end
15
16    | boundTo(name,(n,ol,depth)::t) = if name=n
17                  then ol else boundTo(name,t);
```

Many languages implement bindings in this way. C++, C, Pascal, Standard ML, and to some extent Java all implement bindings in this fashion. Generally, a statically typed language is more likely to be implemented in such a way that bindings are a compile-time entity and don't appear in the target program.

☞ Practice 6.5

What is the list of bindings in the body of the let expression presented in example 6.12?

Not every language is implemented in this way. For example, in chapter 4 the implementation of polymorphism in Ruby is dependent on a run-time lookup of bindings in a hash table. This is true for both function and value bindings. Run-time lookup of bindings has its advantages and disadvantages as described in that chapter. In general though, the earlier an error in a program can be found the more reliable the program will be. At the same time, dynamic binding as implemented in Ruby, gives the programmer a lot of flexibility in the way he or she writes code. The trade-off of flexibility versus safety is one of the fundamental conflicts language designers deal with.

6.8 Defining Scope in Block Structured Languages

The previous section described how let expressions can be added to the calculator language. Let expressions allow the programmer to define identifiers with limited scope. Upon completing the code generation to implement let expressions in the compiler it should be clear how scope is defined.

The scope of an identifier is where that identifier is bound to a value or location in the target program. In the case of the calculator language presented in this chapter, the binding is of an identifier to a location on the run-time stack.

Because the scope of variables in the extended calculator language can be determined at compile-time, the bindings don't have to appear in the compiled program. In other words, in a statically scoped language bindings don't have to be passed around and therefore can be eliminated in the target program.

It is interesting to examine scope as it relates to the visibility of identifiers. While more than one value may be bound to an identifier, only one (identifier, value) pair is visible at a time. After completing the extension of the language to include let expressions it should be clear why this is the case. It has to do with the list of bindings that are passed around in the code generator.

☞ Practice 6.6

Consider the following program.

```
1   let val x = 5
2   in
3     let val y = 10
4     in
5       let val x = 7
6       in
7         x + y
8       end
9       +x
10      end
11  end
```

Label the program by showing all bindings (both visible and invisible) that exist at all appropriate points in the program.
What is the result of executing this program?

This example shows that the scope of a binding determines where a bound value can be accessed. However, not all bindings are necessarily visible at all points in a program. Visibility may be limited by binding two or more values to the same identifier. Many programming languages include the ability to bind the same identifier to multiple values. However, binding an identifier multiple times is not the same as updating a variable with multiple values. Bindings can be undone at a later time while variable assignment is permanent.

6.9 If-Then-Else Expressions

If-then-else expressions allow expressions to be calculated dependent on some condition. The ability to select from one of two choices is often called *selection* in programming language circles. Selection isn't very interesting unless the language supports some sort of interactive input. See example 6.11 for a description of how to add this functionality to the calculator language. The general format of an if-then-else expression is

```
if {Expr} {RelOp} {Expr} then {Expr} else {Expr}
```

where a RelOp is one of $>$, $<$. $>=$, $<=$, $=$, and $<>$. To implement this it is perhaps easiest to define a RelOp token that includes a string with the relational operator's lexeme in it (i.e. like an identifier token). The complete implementation requires changes to the scanner, parser, AST definition, and the code generator. To change the parser a new production should be added to the Factor rules resembling the general format for if-then-else expressions given above.

Code generation is probably the most difficult part of adding selection to the language, but isn't too hard. An example will help in understanding how to generate code for these expressions. Code generation is much simpler if the opposite of the relational operator in the source is used in the target program. For instance, if checking that x $>$ y then generate code to see if x $<=$ y. This will be much clearer in the examples below.

Example 6.17

Here is a program that prints the maximum of two numbers.

```
1  let val x = get
2  in
3    let val y = get
4    in
5      if x > y then x else y
6    end
7  end
```

Code generation begins by generating code for the two expressions in the relational expression. The results of the two expressions are left on the register stack. The two results are popped off and used in the if-then expression that appears below. Two labels are needed in the generated code. Typically, a code generator will have a function called nextLabel that returns a new label each time it is called.

Example 6.18

Here is the target EWE code for the program in example 6.17.

```
1   SP:=100
2   readInt(R0)
3   M[SP+0]:=R0
4   readInt(R1)
5   M[SP+1]:=R1
6   R2:=SP
7   R2:=M[R2+0]
8   R3:=SP
9   R3:=M[R3+1]
10  if R2<=R3 then goto L0
11  R4:=SP
```

```
12   R4:=M[R4+0]
13   goto L1
14   L0:
15   R5:=SP
16   R5:=M[R5+1]
17   L1:
18   writeInt(R5)
19   halt
20
21   equ MEM M[12]
22   equ SP M[13]
23   equ R0 M[0]
24   equ R1 M[0]
25   equ R2 M[0]
26   equ R3 M[1]
27   equ R4 M[0]
28   equ R5 M[0]
```

The label L0 marks the start of the *else* part of the expression. The label L1 marks the end of the if-then expression. The opposite of the relational operator is used in the target code so if the condition is false then the code will go to the *else* clause. Otherwise, the *then* clause is executed. In line 13 (the end of the *then* clause), the code jumps past the *else* clause to the end of the if-then expression so only the *then* or *else* code is executed, but not both.

There is one subtle problem in if-then-else expression code generation. Where is the resulting value when the expression is evaluated? Because register allocation is done at compile-time, knowing which register will hold the result when the program executes is not possible at compile-time. Presumably the choice depends on some input that will not be provided until run-time. However, due to the way register allocation is implemented, it is safe to assume that the *else* clause result register will always hold the result. This is due to the fact that registers that are pushed and popped from the register stack are symbolic registers. Symbolic registers occupy the same physical memory location when possible. Both the *then* clause and *else* clause results are physically stored in the same memory location. In example 6.18 the *then* clause result is in register R4 and the *else* clause result is in register R5. But, both R4 and R5 occupy memory location M[0] so using either register name as the result of the whole if-then expression is safe. In example 6.18 line 18, R5 is used as the if-then expression result.

6.10 Functions in a Block-Structured Language

Implementing a language with functions poses some interesting problems. Functions have two aspects. They are first *implemented* and then *invoked*. Implementing a compiler for a language with functions means that code must be generated for the function implementation and the function invocation.

Block structured languages like Pascal and Standard ML allow the programmer to create local bindings with limited scope. In SML, let expressions are used to define local bindings. However, when local bindings include function declarations some new challenges must be overcome. Section 1.2 describes how an activation record exists for each function invocation. When a function is called an activation record is pushed onto the run-time stack. When a function returns, its activation record is popped from the stack. The primary problem concerns what happens when a function, which has its own activation record, tries to access a value that is in its scope, but bound outside the function's body. That value exists in a different activation record. The example below will help make more sense out of this.

Example 6.19

Consider this program that computes the factorial of n. The base case of the function accesses a value bound to the identifier called base which is not local to the function.

```
1  let val base = 1
2  in
3     let fun fact(n) =
4        if n = 0 then base else n * fact(n-1)
5     in
6        fact(get)
7     end
8  end
```

To keep things simple let's assume that all functions are functions of one parameter. Then, the general format of function definition is

```
let fun id(id) = {Expr} in {Expr} end
```

A function call is denoted by id(Expr). Adding function definitions and function calls to the language greatly increases its power.

Let's assume that 5 is entered at the keyboard. When the function in example 6.19 executes it will push one activation record for the main invocation, and one activation record for each invocation of the fact function. That function will be invoked 6 times. On the sixth invocation the function will need to access the value bound to base which is all the way back in the first activation record. Figure 6.2 depicts the run-time stack at its deepest point while computing fact(5).

The problem comes from the fact that the stack pointer (SP) points at the current activation record and bound values are usually referenced relative to the stack pointer. But, the location bound to base is not in that activation record. In fact, the distance to the correct activation record is unknown since fact(3) might be computed by the program instead of fact(5). There is nothing in the program or data that prevents the fact function from being called with different values of n. If called with a different value, then base is a different distance from the stack pointer. More information is needed if base is to be found.

Block structured languages are languages where the declaration of local variables, constant bindings, and functions may be arbitrarily nested in a block. Blocks

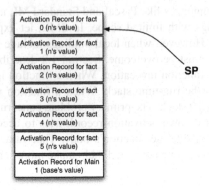

Fig. 6.2: The Run-time Stack

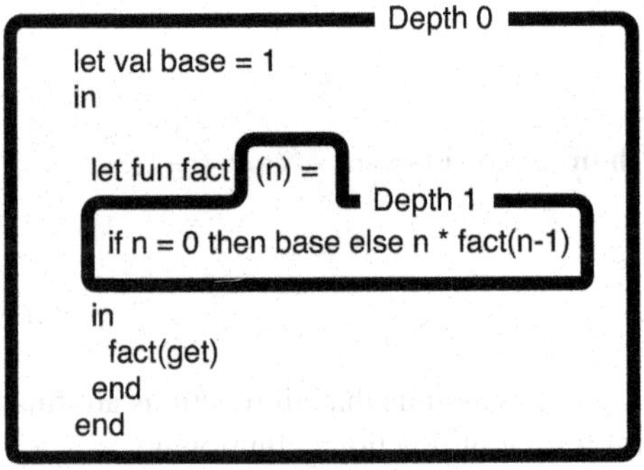

```
┌─────────────────────── Depth 0 ───────────────┐
│                                                │
│   let val base = 1                             │
│   in                                           │
│                                                │
│      let fun fact (n) =                        │
│                      └─ Depth 1 ──────────────┐│
│        if n = 0 then base else n * fact(n-1)   ││
│                                                ─┘
│      in                                        │
│         fact(get)                              │
│      end                                       │
│   end                                          │
└────────────────────────────────────────────────┘
```

Fig. 6.3: Depth Changes in a Program

are delimited in varying ways depending on the language. C uses { ... } for blocks. Pascal uses begin ... end. SML uses let ... in ... end. Languages like C, C++, and Java don't allow nested function declarations. These languages are not truly block structured according to the definition of block structured languages used in this text. Some might argue otherwise. Languages like C, C++ and Java do allow local variables declarations in functions, but since nested function declaration is not allowed to arbitrary depths, the implementation of those languages is simplified while limiting the expressiveness of the language. It should be noted that while the ANSI Standard definition of C does not support nested functions, some implementations do support them as an extension to the language. For instance, the GNU C compiler does support them.

Languages like Algol, Pascal, and SML support nested function declaration. These languages must have additional information available to them to be able to find non-local variables or constant values. One method of providing this informa-

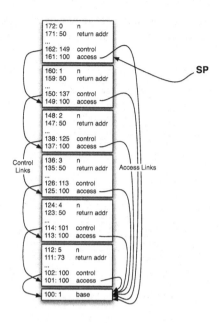

Fig. 6.4: The Run-time Stack with Access and Control Links

tion is to keep track of access links, the depth of each bound identifier, and the current depth. The current depth is depicted in figure 6.3. The depth changes when a function is declared. At the outermost scope the depth is 0. When a function is declared it remains at the depth of the outer scope. The parameter to the function and all local variables or constant values declared inside the function take on the next deeper level's depth. In figure 6.3, the body of the `fact` function is at depth 1. The `base` identifier, declared at depth 0, is assigned a depth of 0. The parameter n is assigned depth 1, the depth of the body of the function where it is declared. The depth of a function's *body* is one more than the depth where the function was declared. The depth of an identifier is the same as the depth where the identifier was declared.

Access links are created in each activation record to enable finding values that are in scope, but not in the current activation record. An access link is a pointer that points to the enclosing block's activation record. The difference between the current depth and the depth of a bound value tells the compiler how many access links must be followed to find a value. For instance, consider the code in figure 6.3. The constant value `base` is declared at depth 0 and is located in the main activation record as depicted in figure 6.4. Recall that an access link points to its enclosing block's activation record. So each call to the `fact` function will have an access link pointing back to the main activation record. When `base` is referenced in the body of `fact` the current depth is 1. The depth of `base` is 0 so the compiler must follow one access link to find the activation record containing `base`'s value. The access link is the first field in each activation record. In any of the calls to fact the access link points to the main activation record, where the value of base is stored.

Offset	Purpose
0	Access Link
1	Control Link
2	Saved PR0
3	Saved PR1
4	Saved PR2
5	Saved PR3
6	Saved PR4
7	Saved PR5
8	Saved PR6
9	Saved PR7
10	Saved PR8 (return address)
11	Saved PR9 (function argument)
12+	Local Variables and Constants

Fig. 6.5: Activation Record Format

The activation record is also used to store a control link. When a function is called an activation record is pushed on the run-time stack. When it returns the activation record is popped. This pushing and popping happens when the value of the stack pointer (SP) is changed. Just before a function returns, the old SP value must be restored to effectively *pop* the activation record. The control link is the old value of the SP. It is saved in the activation record by the caller of the function so it can be restored later just before the function returns.

Register allocation works great in the absence of function calls. But, when a function is invoked the registers that were in use are unknown in the function's code. A function can't know from where it was called so it can't figure out what registers were being used. When code is generated for the body of a function the compiler doesn't know which registers can be used and which cannot. To solve this problem the function must save the contents of all registers it might use. To keep things simple, in this text all registers are saved when a function is called. The ten locations immediately following the control link are used to save the registers of the EWE machine. The symbolic registers of the register allocation framework are all mapped to memory locations M[0] to M[9]. The identifiers PR0 through PR9 can be equated to these memory locations. For instance, PR0 represents physical register 0 which is always equated to M[0]. In this way, a function can save the contents of PR0 to PR9 when the function is called. Then, before the function returns, the physical registers are restored. When the function returns, the code that called it won't know that registers were used by the body of the function. The contents of the registers will have been preserved across the function call.

When a function call is made the program counter is changed to point at the first instruction of the function. When a function returns it must return to wherever it was called. Since a function doesn't know where it was called from, this information must be stored someplace by the calling code so the function can find it. The return address is passed to the function in physical register eight so the function can find it. In addition, the argument's value must also be stored someplace where the function can find it. It is stored in physical register nine by convention. Figure 6.5 shows physical registers PR8 and PR9 are used for these two purposes.

Call/Return Conventions

In any language implementation, function calls only work if the calling and called code cooperate. These rules of cooperation are often called call/return conventions. Some conventions are required to always be the same. But, not every language implementation uses all the same conventions. In fact, even different implementations of the same language may use different conventions. Often the call/return conventions are affected by the architecture the compiled programs will run on. In the case study being examined in this chapter the following conventions are followed. In the conventions below the calling code is the code that is calling the function. The called code is the body of the function.

1. The access link is set by the calling code. This means the new activation record is accessed by the calling code. The next activation record starts at the next available offset after the current activation record.
2. The current SP is stored as the control link by the calling code. This value is stored in the called function's activation record. The location to store the SP register is at the next available offset plus one relative to the current activation record.
3. The argument is stored in PR9 by the calling code during a function call.
4. The Stack pointer is incremented by the calling code to push an activation record on the stack.
5. The return address is stored in PR8 by the calling code.
6. The function is called by the calling code.
7. The called function immediately stores the contents of the registers PR0 to PR9 in its activation record to preserve them across function calls.
8. After generating the code for the body of the function, the result of the body is stored in PR9 so the calling code can find it after the function has returned.
9. The compiler must generate code to restore the registers PR0 to PR8 to their original values before returning from the function.
10. The contents of the stack pointer is restored by the called function to the value stored in the control link field of the activation record.
11. The function returns by restoring the program counter (PC) to the return address (located in PR8).
12. The calling code then places the result of the function call, located in PR9 in the symbolic register designated to hold the result of the function call.

An example of these conventions applied to the factorial program is given in appendix E. The example is labeled with comments to make the code easier to understand.

Setting Access Links

When a function is called, the access link must be set in the next activation record before the call to the function. The access link is set to point to the enclosing block's activation record. When `fact` is called from the main expression the access link points to the current activation record. So the code

```
PR8:=SP                   # set the access link
M[SP+1]:=PR8              # save the access link
```

is generated. However, when `fact` is called recursively then the following code is generated to set the access link.

```
PR8:=SP                   # set the access link
PR8:=M[PR8+0]             # follow the access link
M[SP+12]:=PR8            # save the access link
```

The code that follows the access link is repeated zero or more times depending on the current depth and the depth of the function. The depth of `fact` is 0. The depth of the main expression is 0 so there are 0-0=0 access links followed to set the access link when `fact` is called from the main expression.

When called recursively, the depth of the body of `fact` is 1 and `fact`'s depth is still 0. The difference 1-0=1 says how many access links are followed before setting the new access link when called recursively.

Calling a Function

The program counter, PC, always points at the next instruction to be executed. At the instruction before the function call the PC will be pointing at the function call instruction. The PC+1 will be the return address. In the factorial example the code in example 6.20 is generated when the function is called.

Example 6.20

This code is generated by the compiler when a function is called.

```
1   PR8:=SP                   # set the access link
2   M[SP+1]:=PR8              # save the access link
3   M[SP+2]:=SP               # save the stack pointer
4   PR9:=R9                   # put the parameter in reg 9
5   PR8:=1                    # increment the stack pointer
6   SP:=SP+PR8
```

```
 7  PR8:=PC+1                 # save the return address
 8  goto L1                   # make the fact function call
 9  R9:=PR9                   # put the function result in the result
10                            # symbolic register
```

PC+1 is stored in PR8 and the next line calls the function by going to the label
that denotes the start of the function. The line after the goto statement is the first
line after the function call. The address of this instruction was stored in PR8 on the
line before the function call. After the function call, symbolic register R9 receives
the result of the function which was returned in physical register PR9.

Non-local Access

When accessing a value that is not local to the current activation record, some number of access links must be followed. The number of access links to follow is given
by the difference of the current depth and the depth of the identifier whose value is
being referenced. In the factorial example, when the value bound to base is referenced in the body of the function the following code is generated.

```
R3:=SP       # point to the base of the current activation record
R3:=M[R3+0]  # follow one access link
R3:=M[R3+0]  # load the value into a symbolic register
```

The first line sets the symbolic register to point to the base of the current activation record. The second line follows one access link, the difference between the
current depth and the depth of base. The third line accesses the value bound to
base.

This new way of referencing values in a program can be used when accessing a
local value as well. In that case, zero access links would be followed. This represents
a change in the way values were accessed before implementing function calls.

6.11 Sequential Execution

Sequential execution is interesting when a language has side effects. Assuming that
the calculator language is extended to allow functions, adding the ability to print
values to the screen is trivial (see appendix E to see how a writeln function can
be added to the language). A sequence of one or more statements can be defined as:

```
ExprSeq : Expr                        (Expr)
        | Expr Semicolon ExprSeq      (seq'(Expr,ExprSeq))
```

Once a sequence of expressions like this is defined the code generation for a
sequence is straightforward. Expression sequences appear in the body of let expressions and within parens. For example, the LParen Expr RParen rule can
be replaced by LParen ExprSeq RParen.

Defining sequences in this way allows the programmer to write a sequence of statements anywhere an expression can be written. Just as in SML you can write a sequence either in the body of a `let` expression or by placing parens around the sequence.

6.12   Exercises

1. The language of regular expressions can be used to define the tokens of a language. Give an example for a regular expression from the chapter and indicate what kind of tokens it represents.
2. What does ML-lex do? What input does it require? What does it produce?
3. What does ML-yacc do? What input does it require? What does it produce?
4. How can an abstract syntax tree be expressed in ML?
5. The code generator presented in this chapter does a postfix traversal of an AST. Write an AST representing the expression 5S+R and write some EWE code that represents the compiled version of this expression. Consider the examples presented in this chapter and how similar code could be used to produce your answer. Write some justification for why you are convinced your answer is correct.
6. Given the code presented in example 6.18, consider generating code for a while loop. A while loop has the following general form

   ```
   while {Expr} RelOp {Expr} do {Expr}
   ```

   What code would be generated for this construct? Be as specific as possible. You may indicate the code generated for the various expressions as `codegen(Expr1)`, `codegen(Expr2)`, and `codegen(Expr3)`.
7. Complete the basic calculator language compiler. Finish the code generation for multiplication, division, unary negation, store, and recall. Consult `calcast.sml` and write the part of the code generation function to handle the AST nodes that do not yet have code generated for them. Write the code incrementally. Do just one operator at a time and test it.
8. Write the code given in example 6.11 to implement the `get` operator in the calculator. Be sure to test your program. When you test it, the generated code will display a question mark each time it is waiting for input.
9. Implement let expressions following the information presented in section 6.7 on page 182. When implementing this code be sure to use the `boundTo` function to look up the binding when required. Remember that when code is generated for an expression, the result is left on the top of the register stack. If that result is not needed, the register must still be popped and deleted using `popReg` and `delReg`.
10. Add if-then-else expressions to the calculator language as described in section 6.9. Make use of the `opposite` function in the code to find the opposite of a relational operator when generating the target code. Recall from that section that the final result of the if-then expression will be in the register pushed on the register stack by the *else* clause, even if the *else* clause code was not executed at run-time. See the last paragraph in section 6.9 if you haven't yet read why this is the case.
11. Implement functions as described in section 6.10 starting on page 188. When writing code to follow access links the `forloop` function in `calc.sml` may come in handy. The function will repeatedly invoke some function a given number of time. To print "hello" 5 times you could write

```
forloop(5,TextIO.output,(TextIO.stdOut,"hello\n"));
```

Think carefully about what the bindings and other parameters to codegen should be each time you call it.

12. Add the ability to print values to the screen by examining the code in appendix E. Then add sequential execution to the calculator language as described in section 6.11.

13. Add assignment statements to the language. To do this, add a reference type to the language. Now bindings can have either constant, function, or reference type. A variable is declared by writing

```
let val id = ref {Expr} in {Expr} end
```

The variable given by the identifier is updated by writing an expression like this

```
id := {Expr}
```

This should be declared as a new type of expression in the grammar, not a factor. The variable is dereferenced by writing the identifier preceded by an exclamation point. If x were declared as an integer variable then to add one to x you would write

```
x := !x + 1
```

14. Assuming that variable assignment in the previous exercise has been implemented it is possible to implement iteration in the language. Implement a while loop. A while loop has the form

```
while {Expr} RelOp {Expr} do {Expr}
```

6.13   Solutions to Practice Problems

These are solutions to the practice problems . You should only consult these answers
after you have tried each of them for yourself first. Practice problems are meant to
help reinforce the material you have just read so make use of them.

Solution to Practice Problem 6.1

The keywords `let`, `val`, `in`, `end`, and the symbol = must be added as tokens.
Identifiers must also be added as a token. The last section of the specification
would look like this.

```
1   %%
2   \n   => (pos := (!pos) + 1; lex());
3   {ws}+  => (lex());
4   "("   => (Tokens.LParen(!pos,!pos));
5   ")"   => (Tokens.RParen(!pos,!pos));
6   "+"   => (Tokens.Plus(!pos,!pos));
7   "*"   => (Tokens.Times(!pos,!pos));
8   "/"   => (Tokens.Div(!pos,!pos));
9   "-"   => (Tokens.Minus(!pos,!pos));
10  "=" => (Tokens.Equals(!pos,!pos));
11  {digit}+  => (Tokens.Int(sval(explode yytext,0),!pos,!pos));
12  {alpha}{alphanumeric}* =>
13     (let val tok = String.implode (List.map (Char.toLower)
14                 (String.explode yytext))
15      in
16        if      tok="s" then Tokens.Store(!pos,!pos)
17        else if tok="r" then Tokens.Recall(!pos,!pos)
18        else if tok = "let" then Tokens.Let(!pos,!pos)
19        else if tok = "val" then Tokens.Val(!pos,!pos)
20        else if tok = "in" then Tokens.In(!pos,!pos)
21        else if tok = "end" then Tokens.End(!pos,!pos)
22        else Tokens.ID(yytext,!pos,!pos)
23     end);
24  .  => (error ("error: bad token "^yytext); lex());
```

Solution to Practice Problem 6.2

You need to add two new AST node types. One node must contain the important
`let` information including the identifier (a string) and the two expressions which
are AST nodes themselves. The other type of node is for referring to the bound
value.

```
| letval' of string * AST * AST
| valref' of string
```

Solution to Practice Problem 6.3

The grammar changes required for let expressions are as follows. The ID is needed when a bound value is referred to in an expression.

```
Factor : ...
       | ID                           (valref'(ID))
       | Let Val ID Equal Expr In Expr End
                                      (letval'(ID, Expr1,Expr2))
```

Solution to Practice Problem 6.4

Code is generated in a postfix fashion in general. The code generation for multiplication and unary negation is similar to addition.

```
 1   | codegen(prod'(t1,t2),outFile,bindings,offset,depth) =
 2     let val _ = codegen(t1,outFile,bindings,offset,depth)
 3         val _ = codegen(t2,outFile,bindings,offset,depth)
 4         val reg2 = popReg()
 5         val reg1 = popReg()
 6     in
 7       TextIO.output(outFile,reg1^":="^reg1^"*"^reg2^"\n");
 8       delReg(reg2);
 9       pushReg(reg1)
10     end
11
12   | codegen(negate'(t1),outFile,bindings,offset,depth) =
13     let val _ = codegen(t1,outFile,bindings,offset,depth)
14         val reg1 = popReg()
15         val reg2 = getReg()
16     in
17       TextIO.output(outFile,reg2 ^ ":= 0\n");
18         (* uses registers, rather than specifying a
19            memory location *)
20       TextIO.output(outFile,reg1^":="^reg2^" - "^reg1^"\n");
21       delReg(reg2);
22       pushReg(reg1)
23     end
```

Solution to Practice Problem 6.5

The bindings in the body of the let are [(constant' ("x"), 0, 0)]. There is only one binding of "x" to offset 0 and depth 0.

Solution to Practice Problem 6.6

```
 1   let val x = 5
 2   in                              bindings = [(constant'("x"),0,0)]
 3     let val y = 10
 4     in                            bindings = [(constant'("y"),1,0),
 5                                                (constant'("x"),0,0)]
 6       let val x = 7
 7       in                          bindings = [(constant'("x"),2,0),
 8                                                (constant'("y"),1,0),
 9                                                (constant'("x"),0,0)]
10         x + y
11       end
12     +x
13     end
14   end
```

Recall that bindings are a triple of identifier (with some type information), offset from the SP (i.e. Stack Pointer), and depth which was described in the section on implementing functions. The first x is bound to SP+0 which supposedly holds the value 5. The y is bound to SP+1 which holds 10. Finally, the second x is bound to SP+2 which hold 7.

In the third version of the bindings the first x is not visible even though it is in the bindings. The scope of the first x extends through the body of the expression but it is not always visible since the innermost scope includes another x binding. The result of executing the program is 22.

6.14 Additional Reading

The case study in this chapter illustrated several features of programming languages. The implementation of functions in block structured languages is perhaps the most difficult of the concepts presented. More information on block structured languages can be found in most programming languages texts. The use of ML-lex and ML-yacc is presented on the web in several formats, but to my knowledge is not documented using such a large case study elsewhere.

The implementation of the register allocation framework presented in this text is described in several papers[18, 19, 20]. The development of this framework evolved over several years and works well in practice while simplifying code generation.

The goal of the chapter is to provide an introduction to language features by studying the implementation of a small toy language. Those wishing to learn more about compiler construction may want to consult a full text on the subject. For instance Aho, Sethi, and Ullman's dragon book[1]. There are many other good texts on compiler writing as well.

Chapter 7

Logic Programming

Imperative programming languages reflect the architecture of the underlying von Neumann stored program computer: Programs update memory locations under the control of instructions. Execution is (for the most part) sequential. Sequential execution is governed by a program counter. Imperative programs are prescriptive. They dictate precisely how a result is to be computed by means of a sequence of statements to be performed by the computer.

Example 7.1

Consider this program using the language developed in chapter 6.

```
let val m = ref 0
    val n = ref 0
in
  read(m);
  read(n);
  while !m >= !n do m:=!m-!n;
  writeln(!m)
end
```

What do we want to know about this program? Are we concerned with a detailed decription of what happens when the computer runs this? Do we want to know what the PC is set to when the program finishes? Are we interested in what is in memory location 13 after the second iteration of the loop? These questions are not ones that need to be answered. They don't tell us anything about what the program does.

Instead, if we want to understand the program we want to be able to describe the relationship between the input and the output. The output is the remainder after dividing the first input value by the second input. If this is what we are really concerned about then why not program by describing relationships rather than prescribing a set of steps. This leads to an alternative approach to programming called *Logic Programming*. In Logic Programming the programmer describes the logical structure of a problem rather than prescribing how a computer is to go about solving it. Languages for Logic Programming are called:

- **Descriptive languages:** Programs are expressed as known facts and logical relationships about a problem. Programmers assert the existence of the desired result

K.D. Lee, *Programming Languages*, DOI: 10.1007/978-0-387-79421-1_7,
© Springer Science+Business Media, LLC 2008

and a logic interpreter then uses the computer to find the desired result by making inferences to prove its existence.

- **Nonprocedural languages:** The programmer states only what is to be accomplished and leaves it to the interpreter to determine how it is to be accomplished.
- **Relational languages:** Desired results are expressed as relations or predicates instead of as functions. Rather than define a function for calculating a square root, the programmer defines a relation, say $sqrt(x,y)$, that is true exactly when $y^2 = x$.

While there are many application specific logic programming languages, there is one language that stands out as a general purpose logic programming language. Prolog is the language that is most commonly associated with logic programming. The model of computation for Prolog is not based on the Von Neumann architecture. It's based on the mechanism in logic called unification. Unification is the process where variables are unified to terms.

This text has explored a variety of languages from the EWE assembly language, to C++ and Ruby, to Standard ML, and now Prolog. These languages explore a continuum from very prescriptive languages to descriptive languages.

- Assembly language is a very prescriptive language, meaning that you must think in terms of the particular machine and solve problems accordingly. Programmers must think in terms of the von Neumann machine stored program computer model.
- C++ and Ruby are high-level languages and hence allows you to think in a more descriptive way about a problem. However, the underlying computational model is still the von Neumann machine.
- ML is a high-level language too, but allows the programmer to think in a mathematical way about a problem. This language gets away from the traditional von Neumann model in some ways.
- Prolog takes the descriptive component of languages to the maximum and allows programmers to write programs based solely on describing relationships.

Prolog was developed in 1972. Alain Colmerauer, Phillipe Roussel, and Robert Kowalski were key players in the development of the Prolog language. It is a surprisingly small language with a lot of power. The Prolog interpreter operates by doing a depth first search of the search space while unifying terms to try to come to a conclusion about a question that the programmer poses to the interpreter. The programmer describes relationships and then asks questions.

This simple model of programming has been used in a wide variety of applications including automated writing of real estate advertisements, an application that writes legal documents in multiple languages, another that analyzes social networks, and a landfill management expert system. This is only a sampling of the many, many applications that have been written using this simple but powerful programming model.

7.1 Getting Started with Prolog

If you don't already have a Prolog interpreter, you will want to download one and install it. There are many versions of Prolog available. Some are free and some are not. The standard free implementation is available at http://www.swi-prolog.org. There are binary distributions available for Windows, Mac OS X, and Linux, so there should be something to suit your needs.

Unlike SML, there is no way to write a program interactively with Prolog. Instead, you write a text file, sometimes called a database, containing a list of facts and predicates. Then you start the Prolog interpreter, consult the file, and ask yes or no questions that the Prolog interpreter tries to prove are true.

To start the Prolog interpreter you type either `pl` or `swipl` depending on your installation of SWI Prolog. To exit the interpreter type a `ctl-d`. A Prolog program is a database of facts and predicates that can be used to establish further relationships among those facts. A predicate is a function that returns true or false.

Example 7.2

Prolog programs describe relationships. A simple example is a database of facts about several people in an extended family and the relationships between them.

```
1  parent(fred, sophusw). parent(fred, lawrence).
2  parent(fred, kenny). parent(fred, esther).
3  parent(inger,sophusw). parent(johnhs, fred).
4  parent(mads,johnhs). parent(lars, johan).
5  parent(johan,sophus). parent(lars,mads).
6  parent(sophusw,gary). parent(sophusw,john).
7  parent(sophusw,bruce). parent(gary, kent).
8  parent(gary, stephen). parent(gary,anne).
9  parent(john,michael). parent(john,michelle).
10 parent(addie,gary). parent(gerry, kent).
11 male(gary). male(fred).
12 male(sophus). male(lawrence).
13 male(kenny). male(esther).
14 male(johnhs). male(mads).
15 male(lars). male(john).
16 male(bruce). male(johan).
17 male(sophusw). male(kent).
18 male(stephen). female(inger).
19 female(anne). female(michelle).
20 female(gerry). female(addie).
21 father(X,Y):-parent(X,Y),male(X).
22 mother(X,Y):-parent(X,Y), female(X).
```

Questions we might ask are (1) Is Gary's father Sophus? (2) Who are Kent's fathers? (3) For who is Lars a father? These questions can all be answered by Prolog given this database.

7.2 Fundamentals

Prolog programs (databases) are composed of facts. Facts describe relationships between terms. Simple terms include numbers and atoms. Atoms are symbols like `sophus` that represent an object in our universe of discourse. Atoms MUST start with a small letter. Numbers start with a digit and include both integers and real numbers. Real numbers are written in scientific notation. For instance, 3.14159e0 or just 3.14159 when the exponent is zero.

A predicate is a function that returns true or false. Predicates are defined in prolog by recording a fact or facts about them. For instance, example 7.2 establishes the fact that Johan was the parent of Sophus. `parent` is a predicate representing a true fact about the relationship of `johan` and `sophus`.

Frequently terms include variables in predicate definitions to establish relationships between groups of objects. A variable starts with a capital letter. Variables are used to establish relationships between classes of objects. For instance, to be a father means that you must be a parent of someone and be male. In example 7.2 the `father` predicate is defined by writing

```
father(X,Y):-parent(X,Y), male(X).
```

which means X is the `father` of Y if X is the `parent` of Y and X is `male`. The symbol `:-` is read as *if* and the comma in the predicate definition is read as *and*. So X is a father of Y *if* X is a parent of Y *and* X is male.

☞ Practice 7.1

What are the terms in example 7.2? What is the difference between an atom and a variable? Give examples of terms, atoms, and variables from example 7.2.

To program in Prolog the programmer first writes a database like the one in example 7.2. Then the programmer consults the database so the Prolog interpreter can internally record the facts that are written there. Once the database has been consulted, questions can be asked about the database. Questions asked of Prolog are limited to yes or no questions that are posed in terms of the predicates in the database. A question posed to Prolog is sometimes called a query.

Example 7.3

To discover if Johan is the father of Sophus you start Prolog using `pl` or `swipl`, then consult the database, and pose the query.

```
$ pl
?- consult('family.prolog').
?- father(johan,sophus).
Yes
?-
```

Queries may also contain variables. If we want to find out who the father of sophus is we can ask that of Prolog by replacing the father position in the predicate with a variable.

Example 7.4

When using a variable in a query Prolog will answer yes or no. If the answer is yes, Prolog will tell us what the value of the variable was when the answer was yes. If there is more than one way for the answer to be yes then typing a semicolon will tell Prolog to look for other values where the query is true.

```
?- father(X, sophus).
X = johan
Yes
?- parent(X,kent).
X = gary ;
X = gerry ;
No
?-
```

The final No is Prolog telling us there are no other ways for parent(X,kent) to be true.

The Prolog Program

Prolog performs something called unification to search for a solution. Unification is simply a list of substitutions of terms for variables. A query of the database is matched with its predicate definition in the database. Terms in the query are matched when a suitable pattern is found among the parameters of a predicate in the database. If the matched predicate is dependent on other predicates being true, then those queries are posed to the Prolog interpreter. This process continues until either Prolog finds that no substitution will satisfy the query or it finds a suitable substitution.

Prolog uses depth first search with backtracking to search for a valid substitution. In its search for truth it will unify variables to terms. Once a valid substitution is found it will report the substitution and wait for input. In example 7.4 the interpreter reports that X = gary is a substitution that makes parent(X,kent) true. Prolog waits until either *return* is pressed or a semicolon is entered. When the semicolon is entered, Prolog undoes the last successful substitution it made and continues searching for another substitution that will satisfy the query. In example 7.4 Prolog reports that X = gerry will satisfy the query as well. Pressing semicolon one more time undoes the X = gerry substitution, Prolog continues its depth first search looking for another substitution, finds none, and reports No indicating that the search has exhausted all possible substitutions.

Unification finds a substitution of terms for variables or variables for terms. Unification is a symmetric operation. It doesn't work in only one direction. This means (among other things) that Prolog predicates can run backwards and forwards. For instance, if you want to know who Kent's dad is you can ask that as easily as who is Gary the father of.

Example 7.5

In the following example we find out that gary is the father of kent. We also find out who gary is the father of.

```
?- father(X,kent).
X = gary ;
No
?- father(gary,X).
X = kent ;
X = stephen ;
X = anne ;
No
```

☞ **Practice 7.2**

Write predicates that define the following relationships.

1. brother
2. sister
3. grandparent
4. grandchild

Depending on how you wrote grandparent and grandchild there might be something to note about these two predicates. Do you see a pattern? Why?

7.3 Lists

Prolog supports lists as a data structure. A list is constructed the same as in ML. A list may be empty which is written as [ ] in Prolog. A non-empty list is constructed from an element and a list. The construction of a list with head, H, and tail, T, is written as[H | T]. So, [1,2,3] can also be written as [1| [2 | [3 | [ ]]]]. The list [a | [ ]] is equivalent to writing [a]. Unlike ML, lists in Prolog do not need to be homogeneous. So [1, hi, 4.3] is a valid Prolog list.

By virtue of the fact that Prolog's algorithm is depth first search combined with unification, Prolog naturally does pattern matching. Not only does [H | T] work to construct a list, it also works to match a list with a variable.

Example 7.6

Append can be written as a relationship between three lists. The result of appending the first two lists is the third argument to the append predicate. The first fact below says appending the empty list to the front of y is just y. The second fact says that appending a list whose first element is H to the front of L2 results in [H|T3] when appending T1 and L2 results in T3.

```
append([],Y,Y).
append([H|T1], L2, [H|T3]) :- append(T1,L2,T3).
```

Try out append both backwards and forwards!

Example 7.7

The definition of append can be used to define a predicate called sublist as follows:

```
sublist(X,Y) :- append(_,X,L), append(L,_,Y).
```

Stated in English this says that x is a sublist of y if you can append something on the front of x to get L and something else on the end of L to get y. The underscore is used in the definition for values we don't care about.

To prove that sublist([1],[1,2]) is true we can use the definition of sublist and append to find a substitution for which the predicate holds. Here is an example of using sublist to prove that [1] is a sublist of [1,2].

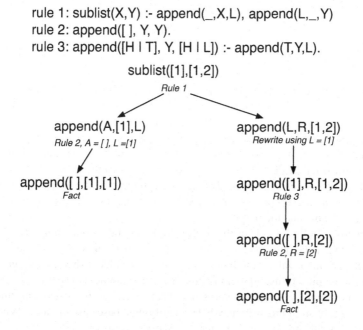

☞ Practice 7.3

What is the complexity of the append predicate? How many steps does it take to append two lists?

☞ Practice 7.4

Write the reverse predicate for lists in Prolog using the append predicate. What is the complexity of this reverse predicate?

The Accumulator Pattern

The slow version of reverse from practice problem 7.4 can be improved upon. The accumulator pattern can be applied to Prolog as it was in SML. Looking back at the solution to practice problem 5.17 on page 165 the ML solution can be rewritten to apply to Prolog as well. In the ML version an accumulator argument was added to the function that allowed the `helprev` helper function to accumulate the reversed list without the use of append.

```
fun reverse(L) =
    let fun helprev (nil, acc) = acc
          | helprev (h::t, acc) = helprev(t,h::acc)
    in
      helprev(L,[])
    end
```

Unlike SML, Prolog does not have any facility for defining local functions with limited scope. If using helper predicates in a Prolog program the user and/or programmer must be trusted to invoke the correct predicates in the correct way.

Applying what we learned from the ML version of reverse to Prolog results in a helprev predicate with an extra argument as well. In many ways this is the same function rewritten in Prolog syntax. The only trick is to remember that you don't write functions in Prolog. Instead, you write predicates. Predicates are just like functions with an extra parameter. The extra parameter establishes the relationship between the input and the output.

Sometimes in Prolog it is useful to think of input and output parameters. For instance, with append defined as a predicate it might be useful to think of the first two parameters as input values and the third as the return value. While as a programmer it might sometimes be useful to think this way, this is not how Prolog works. As was shown in example 7.6, append works both backwards and forwards. But, thinking about the problem in this way may help identifying a base case or cases. When the base cases are identified, the problem may be easier to solve.

☞ Practice 7.5

Write the reverse predicate using a helper predicate to make a linear time reverse using the accumulator pattern.

7.4 Built-in Predicates

Prolog offers a few built in predicates. The relational operators ($<$, $>$, $<=$, $>=$, and $\=$) all works on numbers and are written in infix form. Notice that not equals is written as $\=$ in Prolog.

To check that a predicate doesn't hold, the not predicate is provided. Preceding any predicate with not insists the predicate returns false. For instance, not (5 > 6) returns true because 5 > 6 returns false.

The atom predicate returns true if the argument is an atom. So atom(sophus) returns true but atom(5) does not. The number predicate returns true if the argument is a number. So number(5) returns true but number(sophus) does not.

7.5 Unification and Arithmetic

The Prolog interpreter does a depth first search of the search space while unifying variables to terms. The primary operation that Prolog carries out is unification. Unification can be represented explicitly in a Prolog program by using the equals (i.e. =) operator. When equals is used, Prolog attempts to unify the terms that appear on each side of the operator. If they can be unified, Prolog reports yes and continues unifying other terms to try to find a substitution that satisfies the query. If no substitution is possible, Prolog will report no.

You might have caught yourself wanting to write something like x=y in some of the practice problems . This is normal, but is the sign of a novice Prolog programmer. Writing x=y in a predicate definition is never necessary. Instead, everywhere y appears in the predicate, write x instead.

Unification has one other little nuance that most new Prolog programmers miss. There is no point in unifying a variable to a term if that variable is used only once in a predicate definition. Unification is all about describing relationships. Unification doesn't mean much when a variable is not used in more than one place in a definition. In terms of imperative programming its kind of like storing a value in a variable and then never using the variable. What's the point? Prolog warns us when we do this by saying

```
Singleton variables: [X]
```

If this happens, look for a variable called x (or whatever the variable name is) that is used only once in a predicate definition and replace it with an underscore (i.e. _).

An underscore indicates the result of unification in that position of a predicate isn't needed by the current computation. Prolog warns you of singleton variables because they are a sign that there may be an error in a predicate definition. If an extra variable exists in a predicate definition it may never be instantiated. If that is the case, the predicate will always fail to find a valid substitution. While singleton variables should be removed from predicate definitions, the message is only a warning and does not mean that the predicate is wrong.

The use of equality for unification and not for assignment statements probably seems a little odd to most imperative programmers. The equals operator is not the assignment operator in Prolog. It is unification. Assignment and unification are different concepts. Writing X = 6*5 in Prolog means that the variable X must be equal to the term 6*5, not 30. The equals operator doesn't do arithmetic in Prolog. Instead, a special Prolog operator called `is` is used. To compute 6*5 and assign the result to the variable X the Prolog programmer writes **X is 6*5** as part of a predicate. Using the `is` operator succeeds when the variable on the left is unbound and the expression on the right doesn't cause an exception when computed. All values on the right side of the `is` predicate must be known for the operation to complete successfully. Arithmetic can only be satisfied in one direction, from left to right. This means that predicates involving arithmetic can only be used in one direction, unlike the append predicate and other predicates that don't involve arithmetic.

☞ Practice 7.6

Write a length predicate that computes the length of a list.

7.6 Input and Output

Prolog programs can read from standard input and write to standard output. Reading input is a side-effect so it can only be satisfied once. Once read, it is impossible to unread something. The most basic predicates for getting input are `get_char(X)` which instantiates X to the next character in the input (whatever it is) and `get(X)` which instantiates X to the next non-whitespace character. The `get_char` predicate instantiates X to the character that was read. The `get` predicate instantiates X to the ASCII code of the next character.

There is also a predicate called `read(X)` which reads the next term from the input. When X is uninstantiated, the next term is read from the input and X is instantiated with its value. If X is already instantiated, the next term is read from the input and Prolog attempts to unify the two terms.

Example 7.8

As a convenience, there are certain libraries that also may be provided with Prolog. The `readln` predicate may be used to read an entire line of terms from the keyboard, instantiating a variable to the list that was read. The `readln` predicate has several arguments to control how the terms are read, but typically it can be used by writing

```
? - readln(L,_,_,_,lowercase).
```

Reading input from the keyboard, no matter which predicate is used, causes Prolog to prompt for the input by printing a | : to the screen. If the `readln` predicate is invoked as shown above, entering the text below will instantiate L to the list as shown.

```
|: + 5 S R
L = [+, 5, s, r] ;
No
?-
```

The `print(X)` predicate will print a term to the screen in Prolog. The value of its argument must be instantiated to print it. Print always succeeds even if the argument is an uninstantiated variable. However, printing an uninstantiated variable results in the name of the variable being printed which is probably not what the programmer wants.

Example 7.9

When a query is made in Prolog, each variable is given a unique name to avoid name collisions with other predicates the query may be dependent on. Prolog assigns these unique names and they start with an underscore character. If an uninstantiated variable is printed, you will see it's Prolog assigned unique name.

```
?- print(X).
_G180
X = _G180 ;
No
```

The `print` predicate is satisfied by unifying the variable with the name of Prolog's internal unique variable name which is almost certainly not what was intended. The `print` predicate should never be invoked with an uninstantiated variable.

7.7 Structures

Prolog terms include numbers, atoms, variables and one other important type of term called a structure. A structure in Prolog is like a datatype in SML. Structures are recursive data structures that are used to model structured data. Computer scientists typically call this kind of structured data a tree because they model recursive, hierarchical data. A structure is written by writing a string of characters preceding a tuple of some number of elements.

Example 7.10

Consider implementing a lookup predicate for a binary search tree in Prolog. A tree may be defined recursively as either `nil` or a `btnode(Val,Left,Right)` where `Val` is the value stored at the node and `Left` and `Right` represent the left and right binary search trees. The recursive definition of a binary search tree says that all values in the left subtree must be less than `Val` and all values in the right subtree must be greater than `Val`. For this example, let's assume that binary search trees don't have duplicate values stored in them.

A typical binary search tree structure might look something like this:

```
btnode(5,
  btnode(3,
    btnode(2, nil, nil),
    btnode(4, nil, nil)),
  btnode(8,
    btnode(7, nil, nil),
    btnode(9, nil,
      btnode(10, nil, nil))))
```

which corresponds to the tree shown graphically here.

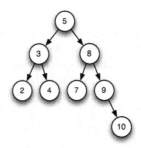

Items may be inserted into and deleted from a binary search tree. Since Prolog programmers write predicates, the code to insert into and delete from a binary search tree must reflect the before and after picture. Because a binary search tree is recursively defined, each part of the definition will be part of a corresponding case for the insert and delete predicates. So, inserting into a search tree involves the value to insert, the tree before it was inserted, and the tree after it was inserted. Similarly, a delete predicate involves the same three arguments.

Looking up a value in a binary search tree results in a true or false response, which is the definition of a predicate. Writing a lookup predicate requires the value and the search tree in which to look for the value.

☞ Practice 7.7

Write a lookup predicate that looks up a value in a binary search tree like the kind defined in example 7.10.

7.8 Parsing in Prolog

As mentioned earlier in the text, Prolog originated out of Colmerauer's interest in using logic to express grammar rules and to formalize the parsing of natural language sentences. Kowalski and Comerauer solved this problem together and Colmerauer figured out how to encode the grammar as predicates so sentences could be parsed efficiently. The next sections describe the implementation of parsing Colmerauer devised in 1972.

Example 7.11

Consider the following context-free grammar for English sentences.

Sentence ::= Subject Predicate .
Subject ::= Determiner Noun
Predicate ::= Verb | Verb Subject
Determiner ::= a | the
Noun ::= professor | home | group
Verb ::= walked | discovered | jailed

Given a sequence of tokens like "the professor discovered a group.", chapter 2 showed that a parse tree can be used to demonstrate that a string is a sentence in the language and at the same time displays its syntactic structure.

☞ Practice 7.8

Construct the parse tree for "the professor discovered a group."

Prolog is especially well suited to parse sentences like the one in practice problem 7.8. The language has built in support for writing grammars and will automatically generate a parser given the grammar of a language. How Prolog does this is not intuitively obvious. The grammar is taken through a series of transformations that produce the parser. The next few pages present these transformations to provide insight into how Prolog generates parsers.

Parsing in Prolog requires the source program, or sentence, be scanned as in the parser implementations presented in chapters 3 and 6. The `readln` predicate discussed on page 212 will suffice to read a sentence from the keyboard and scan the tokens in it. Using the `readln` predicate to read the sentence, "the professor discovered a group.", produces the list [the, professor, discovered, a, group,'.'].

A Prolog parser is a top-down or recursive-descent parser. Because the constructed parser is top-down, the grammar must be LL(1). There cannot be any left-recursive productions in the grammar. Also, because Prolog uses backtracking, there cannot be any productions in the grammar with common prefixes. If there are any common prefixes, left factorization must be performed. Fortunately, the grammar presented in example 7.11 is already LL(1).

The Prolog parser will take the list of tokens and produce a Prolog structure. The structure is the Prolog representation of the abstract syntax tree of the sentence.

For instance, the sentence, "the professor discovered a group.", when parsed by Prolog, yields the term sen(sub(det(the), noun(professor)), pred(verb(discovered), sub(det(a), noun(group)))).

The logic programming approach to analyzing a sentence in a grammar can be viewed in terms of a graph whose edges are labeled by the tokens or terminals in the language.

Example 7.12

This is a graph representation of a sentence. Two terminals are contiguous in the original string if they share a common node in the graph.

O– the →O– professor →O– discovered →O– a →O– group →O– . →O

A sequence of contiguous labels constitutes a nonterminal if the sequence corresponds to the right-hand side of a production rule in the grammar. The contiguous sequence may then be labeled with the nonterminal. In the diagram below three nonterminals are identified.

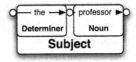

To facilitate the representation of this graph in Prolog the nodes of the graph are given labels. Positive integers are convenient labels to use.

O– the →O– professor →O– discovered →O– a →O– group →O– . →O
1   2   3   4   5   6   7

The graph for the sentence can be represented in Prolog by entering the following facts. These predicates reflect the end points of their corresponding labeled edge in the graph.

```
the(1,2).
professor(2,3).
discovered(3,4).
a(4,5).
group(5,6).
period(6,7).
```

Using the labeled graph above, nonterminals in the grammar can be represented by predicates. For instance, the subject of a sentence can be represented by a subject predicate. The subject(K,L) predicate means that the path from node K to node L can be interpreted as an instance of the subject nonterminal.

For example, subject(4,6) should return true because edge (4,5) is labeled by a determiner "a" and edge (5,6) is labeled by the noun "group". To define a sentence predicate there must exist a determiner and a noun. The rule for the sentence predicate is

```
subject(K,L) :- determiner(K,M), noun(M,L).
```

The common variable M insures the determiner immediately precedes the noun.

☞ Practice 7.9

Construct the predicates for the rest of the grammar.

Example 7.13

The syntactic correctness of the sentence, "the professor discovered a group." can be determined by either of the following queries

```
?- sentence(1,7).
yes
? - sentence(X,Y).
X = 1
Y = 7
```

The sentence is recognized by the parser when the paths in the graph corresponding to the nonterminals in the grammar are verified. If eventually a path for the sentence nonterminal is found then the sentence is valid.

Example 7.14

These are the paths in the graph of the sentence.

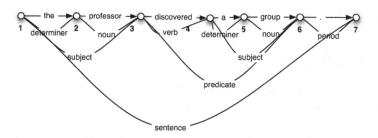

Note the similarity of the structure exhibited by the paths in the graph with the tree of the sentence.

Difference Lists

There are a couple of problems with the development of the parser above. First, entering the sentence as facts like `the(1,2)` and `professor(2,3)` is impractical and awkward. There would have to be some preprocessing on the list to get it in the correct format to be parsed. While this could be done, a better solution exists. The other problem concerns what the parser does. So far the parser only recognizes a syntactically valid sentence and does not produce a representation of the abstract syntax tree for the sentence.

Labeling the nodes of the graph above with integers was an arbitrary decision. The only requirement of labeling nodes in the graph requires that it be obvious when two nodes in the graph are connected. Both problems above can be solved by letting sublists of the sentence label the graph instead of labeling the nodes with integers. These sublists are called difference lists. A difference list represents the part of the sentence that is left to be parsed. The difference between two adjacent nodes is the term which labels the intervening edge.

Example 7.15

This is the difference list representation of the graph.

Using difference lists, two nodes are connected if their difference lists differ by only one element. This connection relationship can be expressed as a Prolog predicate.

Example 7.16

This is the connect predicate and the grammar rewritten to use the connect predicate.

```
c([H|T],H,T).
```

The c (i.e. connect) predicate says that the node labeled `[H|T]` is connected to the node labeled `T` and the edge connecting the two nodes is labeled `H`. This predicate can be used for the terminals in the grammar in place of the facts given above.

```
determiner(K,L) :- c(K,a,L).
determiner(K,L) :- c(K,the,L).

noun(K,L) :- c(K,professor,L).
noun(K,L) :- c(K,home,L).
```

```
noun(K,L)  :- c(K,group,L).

verb(K,L)  :- c(K,walked,L).
verb(K,L)  :- c(K,discovered,L).
verb(K,L)  :- c(K,jailed,L).
```

The graph need not be explicitly created when this representation is employed. The syntactic correctness of the sentence, "the professor discovered a group." can be recognized by the following query.

```
?- sentence([the,professor,discovered,a,group,'.'], [ ]).
yes
```

The parsing succeeds because the node labeled with [the, professor, discovered, a, group, '.'] can be joined to the node labeled with [ ] via the intermediate nodes involved in the recursive descent parse of the sentence. Because Prolog predicates work backwards as well as forward, it is just as easy to explore all the sentences of this grammar by posing this query to the Prolog interpreter.

```
?- sentence(S,[ ]).
```

This reveals that there are 126 different sentences defined by the grammar. Some of the sentences are pretty non-sensical like "the group discovered a group.". Some of the sentences like "the group jailed the professor." have some truth to them. Sophus Lie used to walk to many of the places he visited partly because he liked to walk and partly because he had little money at the time. He also liked to draw sketches of the countryside when hiking. He was jailed in France when France and Germany were at war because the French thought he was a German spy. It was understandable since he was walking through the countryside talking to himself in Norwegian (which the French thought might be German). When they stopped to question him, they found his notebook full of Mathematical formulas and sketchings of the French countryside. He spent a month in prison until they let him go. While in prison he read and worked on his research in Geometry. Of his prison stay he later commented, "I think that a Mathematician is comparatively well suited to be in Prison."[33]. Other mathematicians may not agree with his assessment of the mathematical personality. At least it was better than being shot, which is what happened to at least a few suspected spies at the time.

Some care must be taken when asking for all sentences of a grammar. If the grammar contained a recursive rule, say

```
Subject ::= Determiner Noun | Determiner Noun ``and'' Subject
```

then the language would allow infinitely many sentences, and the sentence generator will get stuck with ever lengthening subject phrases.

7.9 Prolog Grammar Rules

Most implementations of Prolog have a preprocessor which translates grammar rules into Prolog predicates that implement a parser of the language defined by the grammar.

Example 7.17

The grammar of the English language example takes the following form as a logic grammar in Prolog:

```
sentence --> subject, predicate, ['.'].
subject --> determiner, noun.
predicate --> verb, subject.
determiner --> [a].
determiner --> [the].
noun --> [professor]; [home]; [group].
verb --> [walked]; [discovered]; [jailed].
```

Note that terminal symbols appear inside brackets exactly as they look in the source text. Since they are Prolog atoms, tokens starting with characters other than lower case letters must be placed within apostrophes. The Prolog interpreter automatically translates these grammar rules into normal Prolog predicates identical to those defining the grammar presented in the previous section.

Building an AST

The grammar given above is transformed by a preprocessor to generate a Prolog parser. However, in its given form the parser will only answer yes or no, indicating the sentence is valid or invalid. Programmers also want an abstract syntax tree if the sentence is valid. The problem of producing an abstract syntax tree as a sentence is parsed can be handled by using parameters in the logic grammar rules.

Predicates defined using Prolog grammar rules may have arguments in addition to the implicit ones created by the preprocessor. These additional arguments are inserted by the translator to precede the implicit arguments.

Example 7.18

For example, the grammar rule

```
sentence(sen(N,P)) --> subject(N), predicate(P), ['.'].
```

will be translated into the Prolog rule

```
sentence(sen(N,P),K,L) :- subject(N,K,M),
                    predicate(P,M,R),c(R,'.',L).
```

A query with a variable representing a tree produces that tree as its answer.

```
?- sentence(Tree, [the,professor,discovered,a,group,'.'],[]).
Tree = sen(sub(det(the),noun(professor)),
            pred(verb(discovered),sub(det(a),noun(group))))
```

☞ Practice 7.10

Write a grammar for the subset of English sentences presented in this text to parse sentences like the one above. Include parameters to build abstract syntax trees like the one above.

Writing an interpreter or compiler in Prolog is relatively simple given the grammar for the language. Once the AST has been generated for an expression in the language the back end of the interpreter or compiler proceeds much like it does in other languages.

7.10 Exercises

1. In these exercises you should work with the relative database presented at the beginning of this chapter.

    a. Write a rule (i.e. predicate) that describes the relationship of a sibling. Then write a query to find out if Anne and Stephen are siblings. Then ask if Stephen and Mike are siblings. What is Prolog's response?

    b. Write a rule that describes the relationship of a brother. Then write a query to find the brothers of SophusW. What is Prolog's response?

    c. Write a rule that describes the relationship of a niece. Then write a query to find all nieces in the database. What is Prolog's response?

    d. Write a predicate that describes the relationship of cousins.

    e. Write a predicate that describes the relationship of distant cousins. Distant cousins are cousins that are cousins of cousins but not cousins. In other words, your cousins are not distant cousins, but second cousins, third cousins, and so on are distant cousins.

2. Write a predicate called odd that returns true if a list has an odd number of elements.

3. Write a predicate that checks to see if a list is a palindrome.

4. Show the substitution required to prove that sublist([a,b],[c,a,b]) is true. Use the definition in example 7.7 and use the same method of proving its true.

5. Write a predicate that computes the factorial of a number.

6. Write a predicate that computes the nth fibonacci number in exponential time complexity.

7. Write a predicate that computes the nth fibonacci number in linear time complexity.

8. Write a predicate that returns true if a third list is the result of zipping two others together. For instance,

    ```
    zipped([1,2,3],[a,b,c],[pair(1,a),pair(2,b),pair(3,c)])
    ```

    should return true since zipping [1,2,3] and [a,b,c] would yield the list of pairs given above.

9. Write a predicate that counts the number of times a specific atom appears in a list.

10. Write a predicate that returns true if a list is three copies of the same sublist. For instance, the predicate should return true if called as

    ```
    threecopies([a, b, c, a, b, c, a, b, c]).
    ```

    It should also return true if it were called like

    ```
    threecopies([a,b,c,d,a,b,c,d,a,b,c,d]).
    ```

7.11   Solutions to Practice Problems

These are solutions to the practice problems . You should only consult these answers after you have tried each of them for yourself first. Practice problems are meant to help reinforce the material you have just read so make use of them.

Solution to Practice Problem 7.1

Terms include atoms and variables. Atoms include sophus, fred, sophusw, kent, johan, mads, etc. Atoms start with a lowercase letter. Variables start with a capital letter and include X and Y from the example.

Solution to Practice Problem 7.2

1. `brother(X,Y)  :- father(Z,X), father(Z,Y), male(X).`
2. `sister(X,Y)  :- father(Z,X), father(Z,Y), female(X).`
3. `grandparent(X,Y)  :- parent(X,Z), parent(Z,Y).`
4. `grandchild(X,Y)  :- grandparent(Y,X).`

Grandparent and grandchild relationships are just the inverse of each other.

Solution to Practice Problem 7.3

The complexity of append is O(n) in the length of the first list.

Solution to Practice Problem 7.4

```
reverse([],[]).
reverse([H|T],L)  :- reverse(T,RT), append(RT,[H],L).
```

This predicate has $O(n^2)$ complexity since append is called n times and append is $O(n)$ complexity.

Solution to Practice Problem 7.5

```
reverseHelp([],Acc,Acc).
reverseHelp([H|T], Acc, L)  :- reverseHelp(T,[H|Acc],L).
reverse(L,R):-reverseHelp(L,[],R).
```

Solution to Practice Problem 7.6

```
len([],0).
len([H|T],N) :- len(T,M), N is M + 1.
```

Solution to Practice Problem 7.7

```
lookup(X,btNode(X,_,_)).
lookup(X,btNode(Val,Left,_)) :- X < Val, lookup(X,Left).
lookup(X,btNode(Val,_,Right)) :- X > Val, lookup(X,Right).
```

Solution to Practice Problem 7.8

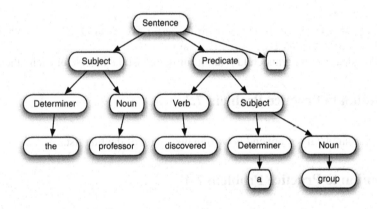

Solution to Practice Problem 7.9

```
sentence(K,L) :- subject(K,M), predicate(M,N), period(N,L).
subject(K,L) :- determiner(K,M), noun(M,L).
predicate(K,L) :- verb(K,M), subject(M,L).
determiner(K,L) :- a(K,L); the(K,L).
verb(K,L) :- discovered(K,L); jailed(K,L); walked(K,L).
noun(K,L) :- professor(K,L); group(K,L); home(K,L).
```

Solution to Practice Problem 7.10

```
sentence(sen(N,P)) --> subject(N), predicate(P), ['.'].
subject(sub(D,N)) --> determiner(D), noun(N).
predicate(pred(V,S)) --> verb(V), subject(S).
determiner(det(the)) --> [the].
determiner(det(a)) --> [a].
noun(noun(professor)) --> [professor].
noun(noun(home)) --> [home].
noun(noun(group)) --> [group].
verb(verb(walked)) --> [walked].
verb(verb(discovered)) --> [discovered].
verb(verb(jailed)) --> [jailed].
```

7.12 Additional Reading

There are several good books on Prolog programming. The Prolog presented in this chapter is enough to get a flavor of the language and a good start programming in the language. Things left out of the discussion include the *cut* operator and some nuances of how unification is done (i.e. the difference between = and ==). Reading from and writing to files was also left out. The definitive book for more information is Clocksin and Mellish[7]. This book lacks exercises but contains many examples and is a good reference once you understand something about how to program in Prolog (which I hope you do now that you've read the chapter and worked through the problems).

Chapter **8**

Formal Semantics

Describing the syntax of a language, as was discussed in chapter 2, is not sufficient for describing the complete meaning of the language. Syntax descriptions say what a program should look like, but not what a program means. The meaning of a language is called the *semantics* of the language by computer scientists. Language designers have long struggled to find the best way to describe the meaning of a language. Most language descriptions rely on English or some other informal language to describe their meaning. Because technical sentences in English can often be interpreted in more than one way, English (or any other informal language) is not necessarily the best choice for a precise definition of a programming language.

Computer scientists have long sought a formal language for describing the meaning of other languages. This area of computer science is referred to as *Formal Semantics of Language Description* or sometimes just *Formal Semantic Methods*. Most formal semantic methods relate the formal syntax of a language to a formal mathematical description of its meaning.

Up to this point the semantics of the languages we have implemented have been described in English and implemented as compilers and interpreters. We have relied on the compiler or interpreter and your own intuitive understanding of programming languages in general to give meaning to the languages you have implemented. This isn't always a good idea.

Why a Formal Method?

For a language to be formally defined means there is no room for interpretation of how programs written in the language should behave. This may seem like an obvious goal of programming languages, but you might be surprised by how differently languages will behave depending on their implementation. Some aspects of a language may be very subtle. Subtle differences may not show up in most programs, but when they do appear it can be difficult to figure out exactly what is going on. This happened to me a few years ago when compiling and running some C++ code.

K.D. Lee, *Programming Languages*, DOI: 10.1007/978-0-387-79421-1_8,
© Springer Science+Business Media, LLC 2008

Example 8.1

Consider this example of C++ code.

```
int x = 1;
cout << x++ << " " << x << endl;
```

The expression x++ is the postfix increment operator. It yields the value stored in x and then increments the x variable. You might expect this code to print "1 2" to the screen. However, the English description of the C++ language, called the Annotated C++ Reference Manual[11], doesn't specify in which order expressions are evaluated within one statement. Therefore, the code could also print "2 1" and that would be just as valid according to the language description. When this code was compiled with two different compilers, the GNU g++ compiled program printed "1 2" while a MIPS C++ compiled version of the program printed "2 1". When you write a program in a modern programming language you expect to get the same results regardless of the compiler or interpreter that you use to compile or run it. Unfortunately, this isn't always the case with programs in many languages.

In another example the language designer tried to be as explicit as possible about the language semantics. But, because English sentences are sometimes ambiguous, different people can read the same text and associate different meanings to it. The originator of Pascal, Niklaus Wirth, described type checking of Pascal programs like this (a paraphrase):

> In Pascal assignment of one variable to another is allowed if the types of the two variables are equivalent.

It appears to be straightforward, but is it?

Example 8.2

This is a very simple Pascal program involving two records that have the same type of fields but different names.

```
program types;
  type A = record
              g:integer;
              h:string;
           end;
       B = record
              i:integer;
              j:string;
           end;
  var x : A;
      y : B;
  begin
    x.g := 5;
    x.h := "hi";
    y := x
  end.
```

Is it a valid program? Well, yes and no. The variables x and y have two different types. One is type A the other type B. A group of grad students under Niklaus Wirth decided that his description of types meant they must be name equivalent. They implemented a compiler that rejected this program. However, Dr. Wirth was fostering competition between his research students and two groups were working on separate compilers for Pascal. The other group read the same description and decided that the program was valid because the types are structurally equivalent. Two different groups of bright people interpreted the same text description in drastically different ways.

Finally, one of the bigger projects in recent years was the development of the Java programming language. Java was designed with the goal of being a cross platform programming environment. This portability hinged on implementations of the Java Virtual Machine for each platform that Java should run on. Early in the development of Java, Sun relied on an English description of the JVM and a very large set of compatibility tests called the JCK (i.e. Java Compatibility Kit). These tests and the English language description of the JVM were supposed to insure that the JVM would work exactly the same independent of the underlying operating system and hardware. The problem was that developers still had trouble implementing the JVM and therefore needed additional help. This help ended up being found in the licensing of the Java Virtual Machine software from Sun. The actual software was needed as a formal description of how the program worked. In November of 2006, Sun announced the release of the JVM source code under the GPL (GNU Public License) to encourage further development of Java and products relating to Java.

The rest of this chapter introduces three methods of formally specifying programming language semantics. This area of Computer Science is still very much a research area. While three methods are presented, none of them are developed well enough to be used by everyone, although Action Semantics has the potential to be used in a wide variety of language descriptions.

8.1 Attribute Grammars

Consider a prefix version of the Calculator expression language first presented in chapter 3. The contents of the memory location after evaluating an expression is not specified by the grammar of the language. In fact, the purpose of any of the operators is not made explicit in the grammar. Even though we know that * stands for multiplication, there is nothing in the grammar itself that insists this be the case. Other means are necessary to convey that meaning. One such method of conveying the semantics of a language is called an *attribute grammar*. An attribute grammar adds attributes to each node of an abstract syntax tree for sentences in the language.

The attributes tell us how a program would be evaluated in terms of its abstract syntax tree. In other words, an attribute grammar provides a mapping of the syntax of a program into a set of attributes that describe the semantics of the program.

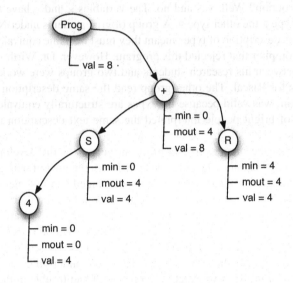

Fig. 8.1: Annotated AST for + S 4 R

Example 8.3

Here is the grammar for prefix calculator expressions. The grammar represents prefix expressions because the operation is written before its arguments. So, + 5 * 6 4 results in 29 when evaluated. Notice that when written in prefix notation, the expression s 5 stores 5 in the memory location. s is now a prefix operator and not a postfix operator as it was previously defined.

$$Prog \rightarrow Expr \ EOF$$
$$Expr \rightarrow op \ Expr \ Expr \ | \ S \ Expr \ | \ number \ | \ R$$
$$where \ op \ is \ one \ of \ +, -, *, /$$

The Prog production in the abstract syntax below was added to assist in the definition of the attribute grammar. Notice that parenthesis have disappeared in the concrete syntax as well as practically all the nonterminals. They aren't needed in a prefix expression grammar. The precedence of operators is determined by the order of operations within the expression. The concrete syntax of the language shown above leads to the abstract syntax description below. Since the precedence of operations is determined by the order they appear, the concrete and abstract syntax are nearly identical.

$$AST \rightarrow Prog \ AST \ | \ op \ AST \ AST \ | \ Store \ AST \ | \ Recall \ | \ number$$
$$where \ op \ is \ one \ of \ +, -, *, /$$

An attribute grammar attaches assignment statements for the attributes to each node in the abstract syntax tree. To distinguish between parts of the abstract syntax tree, let AST_0 denote the AST on the left hand side of a production and AST_i where $i>0$ represent an AST on the right hand side of the production. The attribute grammar for the calculator language is given in example 8.4. Semantics rules are attached to each of the productions in the AST's grammar. These rules govern the assignment of the attributes in the AST. The small numbers to the left of each rule are there simply to number the rules and are not part of the attribute grammar. By deriving an AST for a sentence and then applying the semantic rules the tree is decorated with attributes that describe the meaning of the sentence, or program, in the language.

Example 8.4

Here is the attribute grammar of calculator expressions.

$$AST \rightarrow Prog\ AST$$
(1) $\quad AST_1.min = 0$
(2) $\quad AST_0.val = AST_1.val$
$$AST \rightarrow op\ AST\ AST$$
(3) $\quad AST_1.min = AST_0.min$
(4) $\quad AST_2.min = AST_1.mout$
(5) $\quad AST_0.mout = AST_2.mout$
(6) $\quad AST_0.val = AST_1.val\ op\ AST_2.val$
$\quad\quad$ where op is one of $+, -, *, /$
$$AST \rightarrow Store\ AST$$
(7) $\quad AST_1.min = AST_0.min$
(8) $\quad AST_0.mout = AST_1.val$
(9) $\quad AST_0.val = AST_1.val$
$$AST \rightarrow Recall$$
(10) $\quad AST_0.val = AST_0.min$
(11) $\quad AST_0.mout = AST_0.min$
$$AST \rightarrow number$$
(12) $\quad AST_0.mout = AST_0.min$
(13) $\quad AST_0.val = number$

The attribute grammar given in example 8.4 can be used to convey the meaning of evaluating an expression like + S 4 R. Figure 8.1 depicts the annotated AST according to the attribute grammar given in example 8.4.

☞ **Practice 8.1**

Justify the annotation of the tree given in figure 8.1 by stating which rule was used in assigning each of the attributes annotating the tree.

Attribute Types

Attributes in an attribute grammar come in two flavors. Some attributes are *inherited* which means they are derived from values that are above or to the left in the AST. Some attributes are *synthesized* meaning they are derived from values that are below or to the right in the tree. The val attribute is a synthesized attribute in the attribute grammar presented in example 8.4.

☞ Practice 8.2

Is the min attribute synthesized or inherited? Is the mout attribute synthesized or inherited?

Attribute grammars work great for small languages. When a language is larger, the number of attributes can grow exponentially, resulting in a very large annotated tree. In addition, attribute grammars don't deal well with things like control flow and values that aren't determined until run-time. There are many aspects of programming languages that are difficult to assign as attributes in an AST. Typically, attribute grammars work well for small interpreted languages with little or no unknown information.

8.2 Axiomatic Semantics

The goal in Axiomatic Semantics is to prove the correctness of code. This goal is accomplished through logic considering preconditions and postconditions of statements within a program. The meaning of a program is given in terms of axioms and rules of inference. Each statement in the language is given an inference rule that describes its meaning.

For instance, the meaning of an assignment statement can be given in terms of

$$\{P_E^V\}V := E\{P\} \tag{8.1}$$

where the meaning of P_E^V is that E is substituted for every free occurrence of V in P. Sequential execution of statements has a rule of inference. It is called the Composition rule:

$$\frac{\{P\}S_1\{Q\} \text{ and } \{Q\}S_2\{R\}}{\{P\}S_1;S_2\{R\}} \tag{8.2}$$

Logic is a very precise language. In logic if P implies Q then something stronger than P also implies Q. This is the rule of strengthening preconditions

$$\frac{P \supset Q \text{ and } \{Q\}S\{R\}}{\{P\}S\{R\}} \tag{8.3}$$

and likewise there is a weakening postconditions rule

$$\frac{\{P\}S\{Q\} \text{ and } Q \supset R}{\{P\}S\{R\}} \tag{8.4}$$

If-then-else statements can be descrribed by the the If-Then-Else rule

$$\frac{\{P \wedge B\}S_1\{Q\} \text{ and } \{P \wedge B\}S_2\{Q\}}{\{P\}if \ B \ then \ S_1 \ else \ S_2\{Q\}} \tag{8.5}$$

and If-then statements by the If-Then rule

$$\frac{\{P \wedge B\}S\{Q\} \text{ and } P \wedge B \supset Q}{\{P\}if \ B \ then \ S\{Q\}} \tag{8.6}$$

Example 8.5

Consider the following Pascal code to find the maximum of three numbers:

```
m:= i;                  //statement S1
if m < j then m:=j;     //statement S2
if m < k then m:=k;     //statement S3
```

Assume we would like to prove that m is equal to the maximum of i, j, and k (i.e. $m \geq i$ and $m \geq j$ and $m \geq k$) after the sequence of statements above is executed. To prove this we will need to use the rules of inference that have been defined and the assignment axiom. The assignment axiom is easier to use if used backwards. Starting with the last statement, S3, the proof begins by working through the statements in reverse order.

```
{ m<k and m>=i and m>=j }
       implies { k>=i and k>=j and k=k }
m:=k;
{ m>=i and m>=j and m=k }
       implies { m>=i and m>=j and m>=k }
```

This was derived by applying inference rules 8.4, 8.2, and 8.3 in that order by working backward through the assignment statement. The precondition of the assignment statement in S3 is now `{ m<k and m>=i and m>=j }`. If the condition in S3 is false (i.e. `m>=k`) then we automatically have

```
{ m>=i and m>=j and m>=k }
```

which is the post condition of the whole if-then statement. Therefore, according to the If-Then rule, inference rule 8.2, we have

```
{ m>=i and m>=j } if m < k then m:=k;
     { m>=i and m>=j and m>=k }
```

Similarily, for S2 if `m<j` we find that

```
{ m<j and m=i } implies { j>=i and j=j }
m:=j;
{ m>=i and m=j } implies { m>=i and m>=j }
```

by rules 8.4, 8.3, and 8.2. In S2, if `m>=j` we get

```
{ m=i and m>=j } implies { m>=i and m>=j }
```

Finally, to finish the proof

```
{ true } implies { i=i } m:=i { m=i }
```

Putting it all together results in the following proof.

```
{ true } implies { i=i and j=j and k=k }
m:=i;
{ m=i and j=j and k=k } implies { m>=i and j=j and k=k }
if m < j then m:= j;
{ m>=i and m>=j and k=k }
if m < k then m := k;
{ m>=i and m>=j and m>=k }
```

Axiomatic Semantics has been used for two purposes. Proving properties of specific programs as was done here is one purpose. The other purpose of Axiomatic Semantics centers on formal language definition. The problems associated with Axiomatic Semantics are due to the complexity of finding a proof in a proof system. Fully automated theorem proving is not possible. Proofs must be constructed by a human being using proof techniques and sometimes intuition into how to solve a problem. Proofs get especially complicated when looping structures and other higher level concepts, like exception handling, are added to a language. Loop constructs require the identification of a loop invariant which typically takes some practice to get good at. The inability to automate the proving of program properties means that Axiomatic Semantics will remain a largely academic effort.

8.3 Action Semantics

Action semantics was first proposed by Peter Mosses [25] with the collaboration of David Watt [37]. Several motivations lie behind the design of action semantics. First and foremost, Mosses and Watt wanted a formal language for specifying programming language semantics that was accessible and useful to a wide range of computer scientists, from programmers to language designers.

Action semantics had its roots in another formal semantic method called denotational semantics, so the two meta-languages share many characteristics. Both denotational semantics and action semantics rely on semantic functions to compositionally map an *abstract syntax tree* representing a program to its semantic equivalent, either a *denotation* or an *action*, respectively. Both denotational semantics and action semantics make use of an environment that holds information about the state of the computation. However, the two formal methods define the environment differently.

The Small Language

The Small language is a subset of ML lacking higher order functions but including some of the imperative features of the language. There are also a couple of non-standard functions for input and output which were added to the language. The complete language and its action semantics are presented in appendix F. Many of these language features and their implementation were discussed in chapter 6. The language includes:

- Variables with assignment
- Iteration - while loops
- Selection - if then else statements
- Functions with zero or more parameters
- Input of ints
- Output of ints and bools

Example 8.6

Here is an example Small program which computes the factorial of a number entered at the keyboard and then prints the result to the screen.

```
let fun fact(x) =
  if x=0 then 1
  else
    (output(x);
     x*fact(x-1))
in
  output(fact(input()))
end
```

Running the program produces the following interaction.

```
? 5
5
4
3
2
1
120
```

The Action of Factorial

The formal semantics of this program is given by an action. An action describes the meaning of a program by describing how three things (the transients, bindings, and storage) are modified by the program. Here is the action for the factorial program.

```
||bind "output" to native abstraction of an action
||[ using the given (integer|truth-value) ][ giving () ]
|before
||bind "input" to native abstraction of an action
||[ using the given () ][ giving an integer ]
hence
||furthermore
|||recursively
||||bind "fact" to closure of the abstraction of
||||||furthermore
||||||bind "x" to the given (integer|truth-value)\#1
|||||thence
|||||||||give (integer|truth-value) bound to "x"
|||||||||or
|||||||||give [(integer|truth-value)]cell bound to "x"
||||||||and then
|||||||||give 0
|||||||then
|||||||give the given ((integer|truth-value)|
|||||||              [(integer|truth-value)]cell)\#1
|||||||is the given ((integer|truth-value)|
|||||||              [(integer|truth-value)]cell)\#2
||||||then
|||||||give 1
||||||else
|||||||||give (integer|truth-value) bound to "x"
|||||||||or
|||||||||give [(integer|truth-value)]cell bound to "x"
||||||||then
|||||||||enact application of the abstraction of an action
|||||||||bound to "output" to the given data
||||||||then
|||||||||||give (integer|truth-value) bound to "x"
|||||||||||or
|||||||||||give [(integer|truth-value)]cell bound to "x"
||||||||||and then
||||||||||||||give (integer|truth-value) bound to "x"
||||||||||||||or
||||||||||||||give [(integer|truth-value)]cell bound to "x"
|||||||||||||and then
|||||||||||||give 1
||||||||||||then
||||||||||||||give the difference of
||||||||||||||(the given integer\#1, the given integer\#2)
|||||||||||then
||||||||||||enact application of the abstraction of an action
||||||||||||bound to "fact" to the given data
||||||||||then
|||||||||give the product of the given data
|hence
|||||complete
||||then
|||||enact application of the abstraction of an action bound to
|||||"input" to the given data
```

```
| | |then
| | | |enact application of the abstraction of an action bound to
| | | |"fact" to the given data
| |then
| | |enact application of the abstraction of an action bound to
| | |"output" to the given data
```

Data and Sorts

Meta-languages such as action semantics manipulate semantic objects, called data, that model the data types of the programming languages they describe. Several *sorts* of data are available in action semantics, including integers, lists, maps, characters, and strings among others. Notice that they were referred to as *sorts* of data, not *types* of data. Action semantics is defined using unified algebras [25]. In this context, *sorts* represent a unified approach to data types. The sort datum contains all individuals. Subsorts of datum include the integer, character, and string sorts. The action above contains sorts called variable, value, cell, and 10. Each of these sorts, except the singleton sort 10, have members that are themselves sorts. For instance, the integer 10 is a singleton sort and is also a subsort of the integer and datum sorts. Sorts that contain more than one individual are called proper sorts. A *type* system requires a distinction be made between elements of a type and the type itself. In a *sort* system there is no distinction; elements of a sort are simply subsorts.

data is a sort that consists of tuples of datum. Tuples in action semantics are flat. They cannot be nested. So, the tuple $(10,(5,3))$ is equivalent to $(10,5,3)$.

There are two operators in action semantics for constructing sorts from other sorts, the join and meet operators. The join operator, written using the vertical bar, is a binary sort operator that constructs a new sort consisting of all individuals of the two sorts. The meet operator, written using an ampersand, is also a binary operator that constructs a new sort consisting of all individuals that were contained in both sorts. For example,

integer | truth-value = {false,true,0,-1,1,-2,2,...}
integer & truth-value = nothing
false | true = truth-value

The vacuous sort nothing is a special sort that represents the sort containing no individuals. Sort operators are strict in nothing. For any sort s, the following equations hold

s & nothing = nothing
s | nothing = s

While action semantics defines sorts like integer, character, string, and data, nothing prevents the user of action semantics from extending it to include whatever sort may be needed for the language he or she is describing. Since action semantics is defined using unified algebras, a new sort may be added to an action semantic definition by providing an algebraic specification of its properties.

The Current Information

Action semantics, as the name implies, was designed to provide a means to define programming languages in terms of *actions*. Actions represent computational entities that may be *performed* to modify the *current information* embodied in the form of *storage*, *bindings*, and *transients*. Every action accepts the current information and possibly generates new information that becomes the current information after the performance of the action. An *action semantic definition* of a programming language describes how the current information is modified as a program is executed. The current information is composed of:

- Transients
  Transients represent intermediate, or short-lived, values that are given when evaluating actions. Transients are represented as data, a tuple of datum. For instance, the primitive action 'give 10' produces the transient tuple (10).
- Bindings
  Bindings, represented by a finite mapping sort called map, record the binding of identifiers to a sort called bindable data. bindable data, called denotable values in denotational semantics, usually includes cells, values, and abstractions representing procedures and functions in programming languages. Bindings are created through the use of primitive actions like 'bind "n" to the given value', which produces a map of the identifier "n" to the bindable value.
- Storage
  Storage represents stable information and is a map of locations, called cells to storable data. Storage is persistent: Once the contents of a cell have been altered, it retains that value until it is changed again. Storage, like bindings, is modified by primitive actions. For instance, the primitive action 'store the given value#2 in the given variable#1' stores the datum 10 in the cell bound to the identifier "m" given the appropriate transients and bindings. Note that the sort variable is just another name for the sort cell in this example.

Yielders

Yielders are evaluated to yield information based on the current information. Yielders can access a value in the current information, either in the transients, bindings, or store, and return the value as a transient. For instance, 'given value#2' gives a new transient value from the current transient tuple's second element, if it is a value. A few yielders are:

- given S

  Assumes that the current transient consists of datum d, which is a subsort of S, and yields d. If d is not a subsort of S, then the yielder yields nothing.
- given S#n

  Assumes that the current transient tuple includes a datum d, a subsort of S, at position n in the tuple. It yields d, if d is a subsort of S and nothing otherwise.
- S bound to id

  Expects that the current bindings includes a mapping of id to d, a subsort of S. If it does, then the yielder yields d. If d is not a subsort of S or there is no binding of id in the bindings, then it yields nothing.
- S stored in C

  Assumes that the current storage maps location C to a storable datum d, where d is a subsort of S, and yields d. Otherwise, it yields nothing.

Primitive Actions

Transients, bindings, and storage make up the current information of an action. In the language of action semantics, primitive actions *give* new transient values, *produce* new bindings, or *alter* storage.

For instance, the primitive action 'give 10' gives the transient tuple (10). Primitive actions may use a yielder to interrogate the current information. For instance, the primitive action 'give the given integer' gives the integer in the current transients. Likewise, the primitive actions 'bind _ to _', and 'store _ in _' produce bindings and alter storage, respectively.

Yielders such as those shown above are used to interrogate the current information. Primitive actions use this information to form new transients and bindings and to alter storage. Yielders themselves do not effect the transients or bindings or alter storage. Primitive actions, used in conjunction with yielders, do the work of giving transients, producing bindings, and altering storage.

Two special actions, fail and complete, are the actions that always fail and complete when performed, respectively. These actions are of little interest except in serving as units in action semantics. The complete action represents the action that gives no transients, produces no bindings, and does not alter storage when performed.

Example 8.7

Here are some examples of primitive actions.

- allocate a cell
  - sets aside a new cell in storage and gives it as a transient
- bind M to 5
  - produces the new binding in the outgoing bindings

- store 6 in the given cell
  - stores 6 by mapping the given cell to the value 6.

Combinators and Facets

Action semantics would not be very interesting if primitive actions could not be combined to produce more complex actions. *Combinators* are used to construct *compound actions* from *subactions*. Examples of combinators are and, then, hence, and furthermore. The current information is not explicitly propagated in action semantics (unlike denotational semantics). Combinators are defined to propagate parts of the current information and are also responsible for controlling the order of performance of their subactions. Action semantics contains a wealth of combinators, each with slightly different semantics. When learning action semantics it is difficult to remember how each combinator propagates the current information.

It is useful to study combinators with respect to their *facets*. There are five facets of action semantics, but only three are considered here. They correspond to the three components of the current information: the *functional* facet for transients, the *declarative* facet for bindings, and the *imperative* facet for storage. Every combinator affects these facets in different ways. Remember that every action accepts current information and generates new information. The same is true for compound actions. A combinator affects how the current information is passed to each of its subactions, and how the information generated by each subaction is combined. Most combinators are binary, combining two subactions into one compound action. For instance, consider the action

```
|give 10
then
|bind "n" to the given value
```

In the functional facet the then combinator propagates the transients given by the first subaction to the transients used by the second subaction. So, in this example, 10 is given as the *outgoing* transient of the first subaction and the second subaction uses the singleton tuple 10 as its *incoming* transient value.

The behavior of the then combinator is described by figure 8.2. Combinator diagrams were first introduced by Slonneger [30] and help in understanding the properties of combinators with respect to each facet. The diagrams show the flow of transients and bindings through a compound action with respect to a specific combinator. Transients flow from top to bottom, while bindings flow from left to right. For instance, the then combinator propagates the incoming transients to A. The transients given by A are passed to B, the second subaction. The transients given by the compound action are the transients given by B. The bindings received by the compound action, A then B, are propagated to each of the subactions. Bindings produced by the compound action are the *merged* bindings produced by subactions

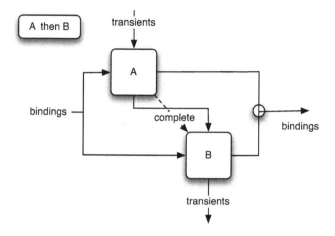

Fig. 8.2: The then Diagram

A and B. The dashed line in figure 8.2 indicates control flow. The then combinator requires that A *completes* first, followed by the performance of B. Since storage is part of the *imperative* facet, changes made to storage by A are seen by B when it is performed.

The verb *complete* also has special meaning in action semantics. Actions may be classified according to their outcomes. Possible outcomes include *completing, failing, diverging, escaping*, and *committing*. So, not only does the dashed line in figure 8.2 indicate control flow, it also indicates that subaction A must *complete* when it is performed.

The behavior of a couple other combinators is shown in figures 8.3 and 8.4. The hence combinator operates like the and combinator on transients and passes the bindings from the first subaction to the second subaction. The unary combinator furthermore overlays the incoming bindings with the bindings produced by subaction A, which is indicated by the broken line in the diagram. furthermore is used by block structured languages to create a new level of scoped identifiers.

The else combinator is a special case. It is used to implement deterministic choice in action semantics. In the action A else B action A is executed if the transients contain true and action B if the transients contain false. The transients, bindings, and storage are modified only by the subaction that is performed.

Mosses showed great insight in the way he defined articles in action semantics, allowing actions to be read more like English. For instance the action 'allocate cell then bind "m" to given variable' does not read nearly as well as 'allocate a cell then bind "m" to the given variable'. Yet, in action semantics both actions have precisely the same meaning because articles, like a and the, are defined as yielders that operate as the identity yielder on data. This definition of articles is convenient and unobtrusive, while making actions much easier to read.

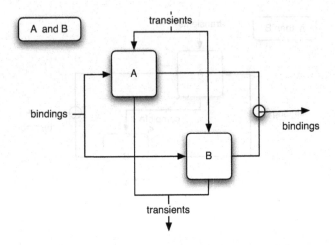

Fig. 8.3: The and Diagram

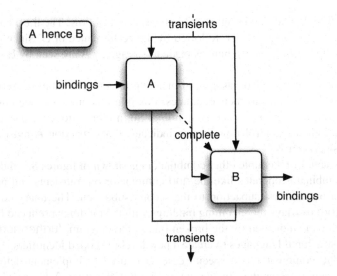

Fig. 8.4: The hence Diagram

Incomes and Outcomes

Recall that actions manipulate the current information (i.e. the transients, bindings, and storage that are provided to an action). It is possible to restrict actions to a subsort of action by the use of incomes and outcomes. Incomes say something about what the current information must contain before performance of an action.

Providing an income or outcome for an action further restricts the sort of the action. For instance, the action below refers to the action that expects to regive an integer. Since it will regive it, it must be given an integer in the first place.

give the given data [using a given integer]

An outcome describes what the current information must look like after the performance an action. So, the action below

give the given data [giving an integer]

is another action that expects to be given an integer and will regive it. Incomes and outcomes may be used to describe any of the facets of action semantics. The incomes and outcomes given above are over the functional facet. The declarative and imperative facets have incomes and outcomes, too. For instance,

bind "m" to the given integer [binding]

is an action that is expected to bind an identifier to something. Over the imperative facet there are outcomes like completing or failing. Incomes and outcomes may also be combined. So, it is possible to describe the action

give the given data [giving an integer] [using an integer] [completing]

Incomes and outcomes can also be applied to yielders and data. It is legal to write

sum (the given integer#1, the given integer#2) [an integer]

for instance. This says that the sum function yields an integer when evaluated.

Action Semantic Descriptions

An Action Semantic Description maps the syntax of a programming language to its action semantics. The mapping is given by a set of semantic functions and equations. From this information it's possible to map an entire source program to its action. Action Semantic Descriptions are modular. It's easy to extend a language definition by adding new semantic equations. Action Semantic Descriptions have the added advantage that they read like a reference manual for a programming language. Action semantics is not only useful to language designers, but also to programmers (once they've learned a little about Action Semantics) and those implementing compilers and interpreters. Action Semantics may be understood on many different levels and therefore used by many different groups of people.

Semantic Functions and Equations

A semantic function is a function from a syntactic category to an action. A syntactic category is just another name for a nonterminal in the grammar. So a semantic description is a mapping from the concrete (or sometimes abstract) syntax of a language to its actions. Consider the evaluation of expressions in the Small language. Expressions may take many forms. To evaluate an expression means to generate the action that corresponds to its evaluation. Here is the declaration of a semantic function. It maps an expression in the Small language to its action.

- evaluate _ :: Expr → action

Semantic functions are *defined* by providing one or more semantic equations that enumerate the possible instances of the semantic function. In the semantic equation below, an *if-then* expression is one possible *Expr* (i.e. expression) in the Small language. The action semantics of an *if-then* expression is given be first evaluating the *SExpr* which yields a boolean value. The boolean value is provided as the transients to the rest of the action which enacts the action generated by the call to *evaluate E_1* if the boolean value is true and enacts the action generated by the call to *evaluate E_2* otherwise. The appearance of *evaluateSExpr* and *evaluate* in the semantic equation are both calls to semantics functions which generate actions when mapping a source program to its action.

(7)   evaluate ⟦ "if" S:SExpr "then" E_1:Expr "else" E_2:Expr ⟧ =
   |evaluateSExpr S
   then
   ||evaluate E_1
   else
   ||evaluate E_2

Sometimes abstract syntax is chosen as the source of the mapping to eliminate extra equations that do little or nothing. However, by using the concrete syntax of a language, the grammar can be extracted directly from the semantic description. More importantly, the semantic description can be given to a tool to generate a compiler from the description. All the necessary information is contained within the description.

Appendix F contains the complete Action Semantic Description of the Small language. Because the description of Small is formally defined as a mapping from the concrete syntax to its action semantics, it is possible to automatically generate a compiler or interpreter from the description. Several projects including Actress[4], Oasis[27], and Genesis[17] have done just that.

8.4 Exercises

1. Using the attribute grammar, construct a decorated abstract syntax tree for the expression + * S 6 5 R. Justify the assignment of attributes by referring to each rule that governed the assignment of a value in the annotated tree.

2. In this exercise you are to construct an interpreter for prefix expressions with the addition of a single memory location (like the interpreter you previously constructed, but implemented in Prolog this time). Your grammar should contain parameters that build an abstract syntax tree of the expression. Then, you should write Prolog rules that evaluate the abstract syntax tree to get the resulting expression. The grammar for prefix expressions is given in example 8.3. The grammar is LL(1) so no modifications of it are necessary to generate a parser for it in Prolog.

   To complete this project you will want to use the readln predicate described in section 7.6. However, to make things easier while parsing, you should preprocess the list so that an expression like "+ S 5 R", which readln returns as [+,s,5,r], will look like [+,s,num(5),r] after preprocessing. The num structure for numbers will help you when you write the parser.

   HINT: This assignment is very closely related to the attribute grammar given in example 8.4. The main predicate for the interpreter should approximate this:

   ```
   calc :- read a line, preprocess the line, parse the expression,
           evaluate the AST, print the result.
   ```

3. Download the Genesis compiler generator and use it to generate the Small compiler. Then use the compiler to compile a Small program. The Small Action Semantic Description is provided with the downloadable Genesis compiler generator. Complete instructions for using Genesis are available on the text's web page.

4. Using Genesis, write an action semantics for the simple calculator expression language. You may use either the prefix version of the grammar given in this chapter or the infix version of the grammar presented in previous chapters. Create the semantic functions that map the concrete syntax of the language directly to their actions. Test your creation to be sure it works. Remember, Genesis is a research project and no guarantees are made regarding appropriate error messages. However, it should be pretty stable as it is written in SML.

8.5 Solutions to Practice Problems

Solution to Practice Problem 8.1

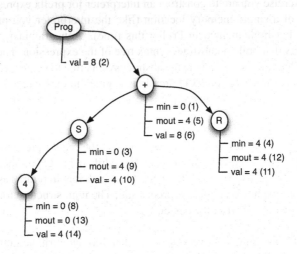

Solution to Practice Problem 8.2

The val attribute is synthesized. The min value is inherited. The mout value is synthesized.

8.6 Additional Reading

Many good resources are available for learning more about formal semantics of programming languages. Slonneger and Kurtz's text[30] is an excellent choice as it covers the topics presented here plus several other formal methods in much more detail than this text. The definitive text for Action Semantics was written by Mosses[26]. Several resources are available on the internet for Action Semantics as well.

Appendix A
The C++ Scanner Class
Implementation

```cpp
1   #include "scanner.h"
2   #include "calcex.h"
3   #include <iostream>
4   #include <string>
5
6   using namespace std;
7
8   //Uncomment this to get debug information
9   //#define debug
10
11  const int numberOfKeywords = 2;
12
13  const string keywd[numberOfKeywords] = {
14    string("S"), string("R")
15  };
16
17  int isLetter(char c) {
18     return ((c >= 'a' && c <= 'z') || (c >= 'A' && c <= 'Z'));
19  }
20
21  int isDigit(char c) {
22    return (c >= '0' && c <= '9');
23  }
24
25  int isWhiteSpace(char c) {
26    return (c == ' ' || c == '\t' || c == '\n');
27  }
28
29  Scanner::Scanner(istream* in):
30    inStream(in),
31    lineCount(1),
32    colCount(-1),
33    needToken(true),
34    lastToken(0)
35  {}
36
37  Scanner::~Scanner() {
38     try {
39        delete inStream;
40     } catch (...) {}
41  }
```

```
42
43   void Scanner::putBackToken() {
44      needToken = false;
45   }
46
47   Token* Scanner::getToken() {
48      if (!needToken) {
49         needToken=true;
50         return lastToken;
51      }
52
53      Token* t;
54      int state=0;
55      bool foundOne=false;
56      char c;
57      string lex;
58      TokenType type;
59      int k;
60      int column, line;
61
62      c = inStream->get();
63
64      while (!foundOne) {
65         colCount++;
66         switch (state) {
67            case 0 :
68               lex = "";
69               column=colCount;
70               line = lineCount;
71               if (isLetter(c)) state=1;
72               else if (isDigit(c)) state=2;
73               else if (c=='+') state=3;
74               else if (c=='-') state=4;
75               else if (c=='*') state=5;
76               else if (c=='/') state=6;
77               else if (c=='(') state=7;
78               else if (c==')') state=8;
79               else if (c=='\n') {
80                  colCount=-1;
81                  lineCount++;
82               }
83               else if (isWhiteSpace(c));
84               else if (inStream->eof()) {
85                  foundOne=true;
86                  type=eof;
87               }
88               else {
89                  cout << "Unrecognized Token found at line " <<
90                     line << " and column " << column << endl;
91                  throw UnrecognizedToken;
92               }
93               break;
94            case 1 :
95               if (isLetter(c) || isDigit(c)) state=1;
```

```
 96              else {
 97                  for (k=0;k<numberOfKeywords;k++)
 98                      if (lex == keywd[k]) {
 99                          foundOne = true;
100                          type = keyword;
101                      }
102                  if (!foundOne) {
103                      type = identifier;
104                      foundOne = true;
105                  }
106              }
107              break;
108          case 2 :
109              if (isDigit(c)) state=2;
110              else {
111                  type = number;
112                  foundOne=true;
113              }
114              break;
115          case 3 :
116              type = add;
117              foundOne = true;
118              break;
119          case 4 :
120              type = sub;
121              foundOne = true;
122              break;
123          case 5 :
124              type = times;
125              foundOne=true;
126              break;
127          case 6 :
128              type = divide;
129              foundOne=true;
130              break;
131          case 7 :
132              type = lparen;
133              foundOne=true;
134              break;
135          case 8 :
136              type = rparen;
137              foundOne=true;
138              break;
139      }

141      if (!foundOne) {
142          lex = lex + c;
143          c = inStream->get();
144      }
145  }

147  inStream->putback(c);
148  colCount--;
149  if (type == number || type == identifier || type == keyword) {
```

```
150        t = new LexicalToken(type, new string(lex), line, column);
151      }
152    else {
153        t = new Token(type, line, column);
154      }
155
156  #ifdef debug
157      cout << "just found " << lex << " with type " << type << endl;
158  #endif
159
160      lastToken = t;
161      return t;
162
163  }
```

Appendix B
The Ruby Scanner Class
Implementation

```ruby
class Scanner
   def initialize(inStream)
      @istream = inStream
      @keywords = Set.new(["S","R"])
      @lineCount = 1
      @colCount = -1
      @needToken = true
      @lastToken = nil
   end

   def putBackToken()
      @needToken = false
   end

   def getToken()
      if !@needToken
         @needToken = true
         return @lastToken
      end

      state = 0
      foundOne = false
      c = @istream.getc()

      if @istream.eof() then
         @lastToken = Token.new(:eof,@lineCount,@colCount)
         return @lastToken
      end

      while !foundOne
         @colCount = @colCount + 1
         case state
         when 0
            lex = ""
            column = @colCount
            line = @lineCount
            if isLetter(c) then state=1
            elsif isDigit(c) then state=2
            elsif c == ?+ then state = 3
            elsif c == ?- then state = 4
            elsif c == ?* then state = 5
```

```
42      elsif c == ?/ then state = 6
43      elsif c == ?( then state = 7
44      elsif c == ?) then state = 8
45      elsif c == ?\n then
46         @colCount = -1
47         @lineCount = @lineCount+1
48      elsif isWhiteSpace(c) then state = state
49         #ignore whitespace
50      elsif @istream.eof() then
51         @foundOne = true
52         type = :eof
53      else
54         puts "Unrecognized Token found at line ",line,
55            " and column ",column,"\n"
56         raise "Unrecognized Token"
57      end
58   when 1
59      if isLetter(c) or isDigit(c) then state = 1
60      else
61         if @keywords.include?(lex) then
62            foundOne = true
63            type = :keyword
64         else
65            foundOne = true
66            type = :identifier
67         end
68      end
69   when 2
70      if isDigit(c) then state = 2
71      else
72         type = :number
73         foundOne = true
74      end
75   when 3
76      type = :add
77      foundOne = true
78   when 4
79      type = :sub
80      foundOne = true
81   when 5
82      type = :times
83      foundOne = true
84   when 6
85      type = :divide
86      foundOne = true
87   when 7
88      type = :lparen
89      foundOne = true
90   when 8
91      type = :rparen
92      foundOne = true
93   end
94
95   if !foundOne then
```

```
96              lex.concat(c)
97              c = @istream.getc()
98          end
99
100      end
101
102      @istream.ungetc(c)
103      @colCount = @colCount - 1
104      if type == :number or type == :identifier or
105          type == :keyword then
106          t = LexicalToken.new(type,lex,line,column)
107      else
108          t = Token.new(type,line,column)
109      end
110
111      @lastToken = t
112      return t
113    end
114
115  private
116    def isLetter(c)
117      return ((?a <= c and c <= ?z) or (?A <= c and c <= ?Z))
118    end
119
120    def isDigit(c)
121      return (?0 <= c and c <= ?9)
122    end
123
124    def isWhiteSpace(c)
125      return (c == ?\  or c == ?\n or c == ?\t)
126    end
127  end
```

Appendix C
Standard ML Basis Library

The Standard ML Basis Library is available at http://standardml.org/Basis in its entirety. This is a subset of the signatures of functions in the basis library. You can use this to quickly look up a function name and its signature. From both the name and the signature you can probably derive the meaning of most functions.

You can also open a structure in the interactive environment of SML to see the contents of it. However, once you open a structure it is open at the top-level and may override some of the functions that are already defined at the top-level. In particular, infix operators on int and real types may no longer be visible at the top-level. Once you open a structure it is probably best to exit and restart the interactive environment.

C.1 The Bool Structure

This is the signature of the functions Bool structure. In addition to the not operator described below, SML defines the *andalso* and *orelse* operators which implement short-circuit logic.

```
1  datatype bool = false | true
2  val not : bool -> bool
3  val toString : bool -> string
4  val fromString : string -> bool option
5  val scan : (char,'a) StringCvt.reader ->
6                          (bool,'a) StringCvt.reader
```

C.2 The Int Structure

This is the signature of the functions in the Int structure.

```
1  type int = ?.int
2  val precision : Int31.int option
3  val minInt : int option
4  val maxInt : int option
5  val toLarge : int -> IntInf.int
6  val fromLarge : IntInf.int -> int
7  val toInt : int -> Int31.int
8  val fromInt : Int31.int -> int
```

```
9    val ~ : int -> int
10   val + : int * int -> int
11   val - : int * int -> int
12   val * : int * int -> int
13   val div : int * int -> int
14   val mod : int * int -> int
15   val quot : int * int -> int
16   val rem : int * int -> int
17   val min : int * int -> int
18   val max : int * int -> int
19   val abs : int -> int
20   val sign : int -> Int31.int
21   val sameSign : int * int -> bool
22   val > : int * int -> bool
23   val >= : int * int -> bool
24   val < : int * int -> bool
25   val <= : int * int -> bool
26   val compare : int * int -> order
27   val toString : int -> string
28   val fromString : string -> int option
29   val scan : StringCvt.radix ->
30        (char,'a) StringCvt.reader -> (int,'a) StringCvt.reader
31   val fmt : StringCvt.radix -> int -> string
```

C.3   The Real Structure

This is the signature of the functions in the Real structure.

```
1    type real = ?.real
2    structure Math :
3      sig
4        type real = real
5        val pi : real
6        val e : real
7        val sqrt : real -> real
8        val sin : real -> real
9        val cos : real -> real
10       val tan : real -> real
11       val asin : real -> real
12       val acos : real -> real
13       val atan : real -> real
14       val atan2 : real * real -> real
15       val exp : real -> real
16       val pow : real * real -> real
17       val ln : real -> real
18       val log10 : real -> real
19       val sinh : real -> real
20       val cosh : real -> real
21       val tanh : real -> real
22     end
23   val radix : int
```

```
24    val precision : int
25    val maxFinite : real
26    val minPos : real
27    val minNormalPos : real
28    val posInf : real
29    val negInf : real
30    val + : real * real -> real
31    val - : real * real -> real
32    val * : real * real -> real
33    val / : real * real -> real
34    val *+ : real * real * real -> real
35    val *- : real * real * real -> real
36    val ~ : real -> real
37    val abs : real -> real
38    val min : real * real -> real
39    val max : real * real -> real
40    val sign : real -> int
41    val signBit : real -> bool
42    val sameSign : real * real -> bool
43    val copySign : real * real -> real
44    val compare : real * real -> order
45    val compareReal : real * real -> IEEEReal.real_order
46    val < : real * real -> bool
47    val <= : real * real -> bool
48    val > : real * real -> bool
49    val >= : real * real -> bool
50    val == : real * real -> bool
51    val != : real * real -> bool
52    val ?= : real * real -> bool
53    val unordered : real * real -> bool
54    val isFinite : real -> bool
55    val isNan : real -> bool
56    val isNormal : real -> bool
57    val class : real -> IEEEReal.float_class
58    val fmt : StringCvt.realfmt -> real -> string
59    val toString : real -> string
60    val fromString : string -> real option
61    val scan : (char,'a) StringCvt.reader ->
62                          (real,'a) StringCvt.reader
63    val toManExp : real -> {exp:int, man:real}
64    val fromManExp : {exp:int, man:real} -> real
65    val split : real -> {frac:real, whole:real}
66    val realMod : real -> real
67    val rem : real * real -> real
68    val nextAfter : real * real -> real
69    val checkFloat : real -> real
70    val floor : real -> int
71    val ceil : real -> int
72    val trunc : real -> int
73    val round : real -> int
74    val realFloor : real -> real
75    val realCeil : real -> real
76    val realTrunc : real -> real
77    val realRound : real -> real
```

```
78    val toInt : IEEEReal.rounding_mode -> real -> int
79    val toLargeInt : IEEEReal.rounding_mode -> real -> IntInf.int
80    val fromInt : int -> real
81    val fromLargeInt : IntInf.int -> real
82    val toLarge : real -> Real64.real
83    val fromLarge : IEEEReal.rounding_mode -> Real64.real -> real
84    val toDecimal : real -> IEEEReal.decimal_approx
85    val fromDecimal : IEEEReal.decimal_approx -> real
```

C.4 The Char Structure

This is the signature of the functions in the Char structure.

```
1     type char = ?.char
2     type string = ?.string
3     val chr : int -> char
4     val ord : char -> int
5     val minChar : char
6     val maxChar : char
7     val maxOrd : int
8     val pred : char -> char
9     val succ : char -> char
10    val < : char * char -> bool
11    val <= : char * char -> bool
12    val > : char * char -> bool
13    val >= : char * char -> bool
14    val compare : char * char -> order
15    val scan : (char,'a) StringCvt.reader ->
16               (char,'a) StringCvt.reader
17    val fromString : String.string -> char option
18    val toString : char -> String.string
19    val fromCString : String.string -> char option
20    val toCString : char -> String.string
21    val contains : string -> char -> bool
22    val notContains : string -> char -> bool
23    val isLower : char -> bool
24    val isUpper : char -> bool
25    val isDigit : char -> bool
26    val isAlpha : char -> bool
27    val isHexDigit : char -> bool
28    val isAlphaNum : char -> bool
29    val isPrint : char -> bool
30    val isSpace : char -> bool
31    val isPunct : char -> bool
32    val isGraph : char -> bool
33    val isCntrl : char -> bool
34    val isAscii : char -> bool
35    val toUpper : char -> char
36    val toLower : char -> char
```

C.5 The String Structure

This is the signature of the functions in the String structure.

```
1    type char = ?.char
2    type string = ?.string
3    val maxSize : int
4    val size : string -> int
5    val sub : string * int -> char
6    val extract : string * int * int option -> string
7    val substring : string * int * int -> string
8    val ^ : string * string -> string
9    val concat : string list -> string
10   val concatWith : string -> string list -> string
11   val str : char -> string
12   val implode : char list -> string
13   val explode : string -> char list
14   val map : (char -> char) -> string -> string
15   val translate : (char -> string) -> string -> string
16   val tokens : (char -> bool) -> string -> string list
17   val fields : (char -> bool) -> string -> string list
18   val isPrefix : string -> string -> bool
19   val isSubstring : string -> string -> bool
20   val isSuffix : string -> string -> bool
21   val compare : string * string -> order
22   val collate : (char * char -> order) ->
23            string * string -> order
24   val < : string * string -> bool
25   val <= : string * string -> bool
26   val > : string * string -> bool
27   val >= : string * string -> bool
28   val fromString : String.string -> string option
29   val toString : string -> String.string
30   val fromCString : String.string -> string option
31   val toCString : string -> String.string
```

C.6 The List Structure

This is the signature of the functions in the List structure.

```
1    datatype 'a list = :: of 'a * 'a list | nil
2    exception Empty
3    val null : 'a list -> bool
4    val hd : 'a list -> 'a
5    val tl : 'a list -> 'a list
6    val last : 'a list -> 'a
7    val getItem : 'a list -> ('a * 'a list) option
8    val nth : 'a list * int -> 'a
9    val take : 'a list * int -> 'a list
10   val drop : 'a list * int -> 'a list
```

```
11   val length : 'a list -> int
12   val rev : 'a list -> 'a list
13   val @ : 'a list * 'a list -> 'a list
14   val concat : 'a list list -> 'a list
15   val revAppend : 'a list * 'a list -> 'a list
16   val app : ('a -> unit) -> 'a list -> unit
17   val map : ('a -> 'b) -> 'a list -> 'b list
18   val mapPartial : ('a -> 'b option) -> 'a list -> 'b list
19   val find : ('a -> bool) -> 'a list -> 'a option
20   val filter : ('a -> bool) -> 'a list -> 'a list
21   val partition : ('a -> bool) -> 'a list -> 'a list * 'a list
22   val foldr : ('a * 'b -> 'b) -> 'b -> 'a list -> 'b
23   val foldl : ('a * 'b -> 'b) -> 'b -> 'a list -> 'b
24   val exists : ('a -> bool) -> 'a list -> bool
25   val all : ('a -> bool) -> 'a list -> bool
26   val tabulate : int * (int -> 'a) -> 'a list
27   val collate : ('a * 'a -> order) -> 'a list * 'a list -> order
```

C.7 The TextIO Structure

This is the signature of the functions in the TextIO structure used for input and output in SML programs.

```
1    type vector = string
2    type elem = char
3    type instream
4    type outstream
5    val input : instream -> vector
6    val input1 : instream -> elem option
7    val inputN : instream * int -> vector
8    val inputAll : instream -> vector
9    val canInput : instream * int -> int option
10   val lookahead : instream -> elem option
11   val closeIn : instream -> unit
12   val endOfStream : instream -> bool
13   val output : outstream * vector -> unit
14   val output1 : outstream * elem -> unit
15   val flushOut : outstream -> unit
16   val closeOut : outstream -> unit
17   structure StreamIO :
18     sig
19       type vector = string
20       type elem = char
21       type reader = reader
22       type writer = writer
23       type instream
24       type outstream
25       type pos = pos
26       type out_pos
27       val input : instream -> vector * instream
28       val input1 : instream -> (elem * instream) option
```

```
29    val inputN : instream * int -> vector * instream
30    val inputAll : instream -> vector * instream
31    val canInput : instream * int -> int option
32    val closeIn : instream -> unit
33    val endOfStream : instream -> bool
34    val mkInstream : reader * vector -> instream
35    val getReader : instream -> reader * vector
36    val filePosIn : instream -> pos
37    val output : outstream * vector -> unit
38    val output1 : outstream * elem -> unit
39    val flushOut : outstream -> unit
40    val closeOut : outstream -> unit
41    val setBufferMode : outstream * buffer_mode -> unit
42    val getBufferMode : outstream -> buffer_mode
43    val mkOutstream : writer * buffer_mode -> outstream
44    val getWriter : outstream -> writer * buffer_mode
45    val getPosOut : outstream -> out_pos
46    val setPosOut : out_pos -> unit
47    val filePosOut : out_pos -> pos
48    val inputLine : instream -> (string * instream) option
49    val outputSubstr : outstream * substring -> unit
50  end
51  val mkInstream : StreamIO.instream -> instream
52  val getInstream : instream -> StreamIO.instream
53  val setInstream : instream * StreamIO.instream -> unit
54  val getPosOut : outstream -> StreamIO.out_pos
55  val setPosOut : outstream * StreamIO.out_pos -> unit
56  val mkOutstream : StreamIO.outstream -> outstream
57  val getOutstream : outstream -> StreamIO.outstream
58  val setOutstream : outstream * StreamIO.outstream -> unit
59  val inputLine : instream -> string option
60  val outputSubstr : outstream * substring -> unit
61  val openIn : string -> instream
62  val openString : string -> instream
63  val openOut : string -> outstream
64  val openAppend : string -> outstream
65  val stdIn : instream
66  val stdOut : outstream
67  val stdErr : outstream
68  val print : string -> unit
69  val scanStream : ((elem,StreamIO.instream) StringCvt.reader
70                      -> ('a,StreamIO.instream) StringCvt.reader)
71                      -> instream -> 'a option
```

Appendix D
SML Calculator Compiler

This is the listing of the calc structure in the calc.sml file.

```sml
structure calc =
struct
open RegisterAllocation;
open calcAS;

    structure calcLrVals =
        calcLrValsFun(structure Token = LrParser.Token)

    structure calcLex =
        calcLexFun(structure Tokens = calcLrVals.Tokens)

    structure calcParser =
        Join(structure Lex= calcLex
                structure LrParser = LrParser
                structure ParserData = calcLrVals.ParserData)

    val input_line =
      fn f =>
        let val sOption = TextIO.inputLine f
        in
            if isSome(sOption) then
                Option.valOf(sOption)
            else
                ""
        end

    val calcparse =
        fn filename =>
          let val instrm = TextIO.openIn filename
                val lexer =
                    calcParser.makeLexer(fn i => input_line instrm)
                val _ = calcLex.UserDeclarations.pos := 1
                val error = fn (e,i:int,_) =>
                        TextIO.output(TextIO.stdOut," line " ^
                        (Int.toString i) ^ ", Error: " ^ e ^ "\n")
          in
          calcParser.parse(30,lexer,error,())
                before TextIO.closeIn instrm
          end

    exception IdNotBound;

    fun lookup(id, nil) =
```

```
44          (TextIO.output (TextIO.stdOut, "Identifier "^id^
45             " not declared!\n");
46          raise IdNotBound)
47     | lookup(id,(x,offset)::L) =
48          if id = x then offset else lookup(id,L)
49
50   fun codegen(add'(t1,t2),outFile,bindings,nextOffset) =
51        let val _ = codegen(t1,outFile,bindings,nextOffset)
52            val _ = codegen(t2,outFile,bindings,nextOffset)
53            val reg2 = popReg()
54            val reg1 = popReg()
55        in
56          TextIO.output (outFile,reg1 ^ ":="^reg1^"+"^reg2^"\n");
57          delReg(reg2);
58          pushReg(reg1);
59          ([],nextOffset)
60        end
61
62     | codegen(sub'(t1,t2),outFile,bindings,nextOffset) =
63        let val _ = codegen(t1,outFile,bindings,nextOffset)
64            val _ = codegen(t2,outFile,bindings,nextOffset)
65            val reg2 = popReg()
66            val reg1 = popReg()
67        in
68          TextIO.output (outFile,reg1 ^ ":="^reg1^"-"^reg2^"\n");
69          delReg(reg2);
70          pushReg(reg1);
71          ([],nextOffset)
72        end
73
74     | codegen(integer'(i),outFile,bindings,nextOffset) =
75        let val r = getReg()
76        in
77          TextIO.output (outFile, r ^ ":=" ^
78             Int.toString(i) ^ "\n");
79          pushReg(r);
80          ([],nextOffset)
81        end
82
83
84   fun compile filename =
85        let val (ast, _) = calcparse filename
86            val outFile = TextIO.openOut("a.ewe")
87        in
88          TextIO.output (outFile,"SP:=100\n");
89          let val s = codegen(ast,outFile,[],0)
90              val reg1 = popReg()
91          in
92            TextIO.output (outFile, "writeInt("^reg1^")\nhalt\n\n");
93            delReg(reg1);
94            TextIO.output (outFile,"equ MEM M[12]\n");
95            printRegs(!regList,outFile);
96            TextIO.output (outFile,"equ SP M[13]\n");
97            TextIO.closeOut(outFile)
```

```
98            end
99          end
100
101    fun run(a,b::c) = (compile b; OS.Process.success)
102      | run(a,b) = (TextIO.print("usage: sml @SMLload=calc\n");
103                    OS.Process.success)
104  end
```

Appendix E
The Factorial Program's Code

Here is the compiled EWE code for the program presented in example 6.19. The code is commented to aid in understanding the call/return conventions presented in the text.

```
 1  SP:=100
 2  PR0 := 0
 3  PR1 := 0
 4  PR2 := 0
 5  PR3 := 0
 6  PR4 := 0
 7  PR5 := 0
 8  PR6 := 0
 9  PR7 := 0
10  PR8 := 0
11  PR9 := 0
12  cr := 13
13  nl := 10
14  nullchar:=0
15  R0:=1
16  M[SP+0]:=R0
17  goto L0                  # jump around the function body
18  ###### function fact(n) ######
19  # function prolog
20  L1:  M[SP+2]:=PR0        # save the registers in the run-time stack
21  M[SP+3]:=PR1
22  M[SP+4]:=PR2
23  M[SP+5]:=PR3
24  M[SP+6]:=PR4
25  M[SP+7]:=PR5
26  M[SP+8]:=PR6
27  M[SP+9]:=PR7
28  M[SP+10]:=PR8
29  M[SP+11]:=PR9
30  # body of the function
31  R1:=SP
32  R1:=M[R1+11]
33  R2:=0
34  if R1<>R2 then goto L2
35  R3:=SP
36  R3:=M[R3+0]
37  R3:=M[R3+0]
38  goto L3
39  L2:
40  R4:=SP
```

```
41   R4:=M[R4+11]
42   R5:=SP
43   R5:=M[R5+11]
44   R6:=1
45   R5:=R5-R6
46   PR8:=SP              # set the access link
47   PR8:=M[PR8+0]        # follow the access link
48   M[SP+12]:=PR8        # save the access link
49   M[SP+13]:=SP         # save the stack pointer
50   PR9:=R5              # put the parameter in reg 9
51   PR8:=12              # increment the stack pointer
52   SP:=SP+PR8
53   PR8:=PC+1            # save the return address
54   goto L1              # make the fact function call
55   R5:=PR9              # put the function result in the
56                        # result symbolic register
57   R4:=R4*R5
58   L3:
59   # end of the function body
60   # function epilog
61   PR9:=R4              # save the function result
62   PR0:=M[SP+2]         # restore the registers
63   PR1:=M[SP+3]
64   PR2:=M[SP+4]
65   PR3:=M[SP+5]
66   PR4:=M[SP+6]
67   PR5:=M[SP+7]
68   PR6:=M[SP+8]
69   PR7:=M[SP+9]
70   PR8:=M[SP+10]
71   SP:=M[SP+1]          # restore the stack pointer
72   PC:=PR8              # return from the function
73   L0:
74   readInt(R9)
75   PR8:=SP              # set the access link
76   M[SP+1]:=PR8         # save the access link
77   M[SP+2]:=SP          # save the stack pointer
78   PR9:=R9              # put the parameter in reg 9
79   PR8:=1               # increment the stack pointer
80   SP:=SP+PR8
81   PR8:=PC+1            # save the return address
82   goto L1              # make the fact function call
83   R9:=PR9              # put the function result in the
84                        # result symbolic register
85   writeInt(R9)
86   halt
87
88   ###### input function ######
89   input:  readInt(PR9)
90                        # read an integer into
91                        # function result register
92   SP:=M[SP+1]          # restore the stack pointer
93   PC:=PR8              # return from whence you came
94   ###### output function ######
```

```
 95  output:  writeInt(PR9)
 96                       # write the integer in function
 97                       # parameter register
 98  writeStr(cr)
 99  SP:=M[SP+1]     # restore the stack pointer
100  PC:=PR8         # return from whence you came
101  equ PR0 M[0]
102  equ PR1 M[1]
103  equ PR2 M[2]
104  equ PR3 M[3]
105  equ PR4 M[4]
106  equ PR5 M[5]
107  equ PR6 M[6]
108  equ PR7 M[7]
109  equ PR8 M[8]
110  equ PR9 M[9]
111  equ MEM M[12]
112  equ SP M[13]
113  equ cr M[14]
114  equ nl M[15]
115  equ nullchar M[16]
116  equ R0 M[0]
117  equ R1 M[0]
118  equ R2 M[1]
119  equ R3 M[0]
120  equ R4 M[0]
121  equ R5 M[1]
122  equ R6 M[2]
123  equ R7 M[2]
124  equ R8 M[3]
125  equ R9 M[0]
126  equ R10 M[1]
127  equ R11 M[2]
```

Appendix F
Small Action Semantic Description

Sorts

(1) function = an abstraction of an action [producing the empty-map]
(2) value = integer | truth-value
(3) bindable = value | [value]cell | function

Semantics

- run _ :: Prog → action
(1) run ⟦ E:Expr ⟧ =
> ‖bind "output" to the native abstraction of an action
> ‖[using a given value] [giving ()]
> before
> ‖bind "input" to the native abstraction of an action
> ‖[giving an integer] [using the given ()]
> hence
> ‖evaluate E .
- evaluateSeq _ :: ExprSeq → action
(2) evaluateSeq ⟦ E:Expr ";" E_s:ExprSeq ⟧ =
> ‖evaluate E
> then
> ‖evaluateSeq E_s
(3) evaluateSeq ⟦ E:Expr ⟧ =
> evaluate E
- evaluate _ :: Expr → action
(4) evaluate ⟦ L_v:Expr ":=" E:Expr ⟧ =
> ‖evaluate L_v
> ‖and then
> ‖evaluate E
> then
> ‖store the given value#2 in the given cell#1

(5)   evaluate ⟦ "while" B:Expr "do" E:Expr ⟧ =
⎸ unfolding
⎸ ⎸evaluate B
⎸ ⎸then
⎸ ⎸⎸⎸evaluate E
⎸ ⎸⎸and then
⎸ ⎸⎸unfold
⎸ ⎸else
⎸ ⎸⎸complete

(6)   evaluate ⟦ "let" D_s:DecSeq "in" E_s:ExprSeq "end" ⟧ =
⎸furthermore elaborateDecSeq D_s
⎸hence
⎸evaluateSeq E_s

(7)   evaluate ⟦ "if" S:SExpr "then" E_1:Expr "else" E_2:Expr ⟧ =
⎸evaluateSExpr S
⎸then
⎸⎸evaluate E_1
⎸else
⎸⎸evaluate E_2

(8)   evaluate ⟦ S:SExpr ⟧ =
⎸evaluateSExpr S

  • elaborateDecSeq _ :: DecSeq → action

(9)   elaborateDecSeq ⟦ D:Decl D_s:DecSeq ⟧ =
⎸elaborate D before elaborateDecSeq D_s

(10)   elaborateDecSeq ⟦ D:Decl ⟧ =
⎸elaborate D

  • elaborate _ :: Decl → action

(11)   elaborate ⟦ "fun" F_s:FunSeq ⟧ =
⎸recursively elaborateFunSeq F_s

(12)   elaborate ⟦ "val" I:Identifier "=" E:Expr ⟧ =
⎸evaluate E
⎸then
⎸bind I to the given (value | [value]cell)

  • elaborateFunSeq _ :: FunSeq → action

(13)   elaborateFunSeq ⟦ F:FunDecl "and" F_s:FunSeq ⟧ =
⎸elaborateFun F and then elaborateFunSeq F_s

(14)   elaborateFunSeq ⟦ F:FunDecl ⟧ =
⎸elaborateFun F

  • elaborateFun _ :: FunDecl → action

(15)   elaborateFun ⟦ Id:Identifier "(" ")" "=" E:Expr ⟧ =
⎸bind I to the closure of the abstraction of
⎸evaluate E

(16) elaborateFun ⟦ *I*:Identifier "(" *P_s*:FormalParmSeq ")" "=" *E*:Expr ⟧ =
    bind *I* to the closure of the abstraction of
      ‖furthermore
      ‖bindParameters *P_s*
     thence
     ‖evaluate *E*

- bindParameters _ :: FormalParmSeq → action

(17) bindParameters ⟦ *I*:Identifier ⟧ =
    bind *I* to the given value

(18) bindParameters ⟦ *I*:Identifier ":" "int" ⟧ =
    bind *I* to the given integer

(19) bindParameters ⟦ *I*:Identifier ":" "bool" ⟧ =
    bind *I* to the given truth-value

(20) bindParameters ⟦ *P*:FormalParm "," *P_s*:FormalParmSeq ⟧ =
    ‖bindParameter *P*
    and then
    ‖give the rest of the given data
   then
   ‖bindParameters *P_s*

- bindParameter _ :: FormalParm → action

(21) bindParameter ⟦ *I*:Identifier ⟧ =
    bind *I* to the given value#1

(22) bindParameter ⟦ *I*:Identifier ":" "int" ⟧ =
    bind *I* to the given integer#1

(23) bindParameter ⟦ *I*:Identifier ":" "bool" ⟧ =
    bind *I* to the given truth-value#1

- evaluateSExpr _ :: SExpr → action

(24) evaluateSExpr ⟦ "ref" *E*:SExpr ⟧ =
    ‖allocate a cell
    and then
    ‖evaluateSExpr *E*
   then
    ‖store the given value#2 in the given cell#1
    and then
    ‖give the given cell#1

(25) evaluateSExpr ⟦ *E*:SExpr "orelse" *T*:Term ⟧ =
    ‖evaluateSExpr *E*
   then
    ‖give true
    else
    ‖evaluateTerm *T*

(26) evaluateSExpr ⟦ *E*:SExpr "+" *T*:Term ⟧ =
    ||evaluateSExpr *E*
    and then
    ||evaluateTerm *T*
    then
    |give the sum of the given data

(27) evaluateSExpr ⟦ *E*:SExpr "-" *T*:Term ⟧ =
    ||evaluateSExpr *E*
    and then
    ||evaluateTerm *T*
    then
    |give the difference of (the given integer#1, the given integer#2)

(28) evaluateSExpr ⟦ *T*:Term ⟧ =
    evaluateTerm *T*

- evaluateTerm _ :: Term → action

(29) evaluateTerm ⟦ *T*:Term "andalso" *N*:Neg ⟧ =
    |evaluateTerm *T*
    then
    ||evaluateNeg *N*
    else
    ||give false

(30) evaluateTerm ⟦ *T*:Term "*" *N*:Neg ⟧ =
    ||evaluateTerm *T*
    and then
    ||evaluateNeg *N*
    then
    |give the product of the given data

(31) evaluateTerm ⟦ *T*:Term "/" *N*:Neg ⟧ =
    ||evaluateTerm *T*
    and then
    ||evaluateNeg *N*
    then
    |give the (quotient of (the given integer#1, the given integer#2))
    |[yielding an integer]

(32) evaluateTerm ⟦ *T*:Term "mod" *N*:Neg ⟧ =
    ||evaluateTerm *T*
    and then
    ||evaluateNeg *N*
    then
    |give remainder of (the given integer#1, the given integer#2)

(33) evaluateTerm ⟦ *N*:Neg ⟧ =
    evaluateNeg *N*

- evaluateNeg _ :: Neg → action

(34) evaluateNeg ⟦ "not" N:Neg ⟧ =
  |evaluateNeg N
  then
  |give not of the given truth-value
(35) evaluateNeg ⟦ "-" N:Neg ⟧ =
  |evaluateNeg N
  then
  |give (difference of (0, the given integer)) [yielding an integer]
(36) evaluateNeg ⟦ C:Comparison ⟧ =
      evaluateComparison C
  • evaluateComparison _ :: Comparison → action
(37) evaluateComparison ⟦ F_1:Factor "<" F_2:Factor ⟧ =
  ||evaluateFactor F_1
  and then
  |evaluateFactor F_2
  then
  |give the given integer#1 is less than the given integer#2
(38) evaluateComparison ⟦ F_1:Factor ">" F_2:Factor ⟧ =
  ||evaluateFactor F_1
  and then
  |evaluateFactor F_2
  then
  |give the given integer#1 is greater than the given integer#2
(39) evaluateComparison ⟦ F_1:Factor "=" F_2:Factor ⟧ =
  ||evaluateFactor F_1
  and then
  |evaluateFactor F_2
  then
  |give the given (value | [value]cell) #1 is
  |the given (value | [value]cell)#2
(40) evaluateComparison ⟦ F_1:Factor "<>" F_2:Factor ⟧ =
  ||evaluateFactor F_1
  and then
  |evaluateFactor F_2
  then
  |give not (the given (value | [value]cell)#1 is
  |the given (value | [value]cell)#2)
(41) evaluateComparison ⟦ F_1:Factor ">=" F_2:Factor ⟧ =
  ||evaluateFactor F_1
  and then
  |evaluateFactor F_2
  then
  |give not (the given integer#1 is less than the given integer#2)

(42) evaluateComparison ⟦ F_1:Factor "<=" F_2:Factor ⟧ =
      ‖evaluateFactor F_1
      and then
      ‖evaluateFactor F_2
      then
      ‖give not (the given integer#1 is greater than the given integer#2)

(43) evaluateComparison ⟦ F:Factor ⟧ =
      evaluateFactor F

• evaluateFactor _ :: Factor → action

(44) evaluateFactor ⟦ I:Integer ⟧ =
      give I

(45) evaluateFactor ⟦ "true" ⟧ =
      give true

(46) evaluateFactor ⟦ "false" ⟧ =
      give false

(47) evaluateFactor ⟦ "!" F:Factor ⟧ =
      ‖evaluateFactor F
      then
      ‖give the value stored in the given cell

(48) evaluateFactor ⟦ I:Identifier ⟧ =
      ‖give the value bound to I
      or
      ‖give the [value]cell bound to I

(49) evaluateFactor ⟦ I:Identifier "(" A:ArgSeq ")" ⟧ =
      ‖evaluateArgs A
      then
      ‖enact the application of the function bound to I to the given data

(50) evaluateFactor ⟦ I:Identifier "(" ")" ⟧ =
      ‖complete
      then
      ‖enact the application of the function bound to I to the given data

(51) evaluateFactor ⟦ "(" S:ExprSeq ")" ⟧ =
      evaluateSeq S

• evaluateArgs _ :: ArgSeq → action

(52) evaluateArgs ⟦ E:Expr ⟧ =
      evaluate E

(53) evaluateArgs ⟦ E:Expr "," A:ArgSeq ⟧ =
      ‖evaluate E
      and then
      ‖evaluateArgs A

References

1. Alfred V. Aho, Ravi Sethi, and Jeffrey D. Ullman. *Compilers: principles, techniques, and tools*. Addison-Wesley Longman Publishing Co., Inc., Boston, MA, USA, 1986.
2. John Backus. *[Photograph]*. Photograph provided courtesy of IBM and used with permission., 2008.
3. Edoardo Biagioni. A structured tcp in standard ml. In *SIGCOMM '94: Proceedings of the conference on Communications architectures, protocols and applications*, pages 36–45, New York, NY, USA, 1994. ACM Press.
4. D.F. Brown, H. Moura, and D.A. Watt. Actress: an action semantics directed compiler generator. In *Proceedings of the Workshop on Compiler Construction*, Paderborn, Germany, 1992.
5. Timothy A. Budd. *C++ for Java Programmers*. Addison-Wesley Publishing Co., Inc., Boston, MA, USA, 1999.
6. L. Cardelli. *Handbook of Computer Science and Engineering*. CRC Press - Digital Equipment Corporation, 1997.
7. W. Clocksin and C. Mellish. *Programming in Prolog*. Springer, 2003.
8. Alain Colmerauer. *[Photograph]*. Photograph provided courtesy of Alain Colmerauer and used with his permission., 2008.
9. Alain Colmerauer and Philippe Roussel. The birth of prolog. In *HOPL-II: The second ACM SIGPLAN conference on History of programming languages*, pages 37–52, New York, NY, USA, 1993. ACM.
10. British Broadcasting Corporation. 248-dimension math puzzle solved, 2007. [Online; accessed 3-March-2008].
11. M. Ellis and B. Stroustrup. *The Annotated C++ Reference Manual*. Addison-Wesley Professional, 1990.
12. R. Girvan. Partial differential equations. *Scientific-Computing.com*, 2006. http://www.scientific-computing.com/review4.html.
13. Mike Gordon. From lcf to hol: a short history. pages 169–185, 2000.
14. Robert Harper and Peter Lee. Advanced languages for systems software: The Fox project in 1994. Technical Report CMU-CS-94-104, School of Computer Science, Carnegie Mellon University, Pittsburgh, PA, January 1994. (Also published as Fox Memorandum CMU-CS-FOX-94-01).
15. Robert Kowalski. An interview with robert kowalski, 2008. Details of events provided by Robert Kowalski through an exchange of email from 12-February-2008 to 14-February-2008.
16. Robert Kowalski. *[Photograph]*. Photograph provided courtesy of Robert Kowalski and used with his permission., 2008.
17. K.D. Lee. *Action Semantics-based Compiler Generation*. PhD thesis, Department of Computer Science, University of Iowa, 1999.
18. K.D. Lee. Minimal register allocation. Technical Report 99-06, University of Iowa, Department of Computer Science, Iowa City, IA, 1999.

19. K.D. Lee and H. Zhang. Formal development of a minimal register allocation algorithm. Technical Report 99-07, University of Iowa, Department of Computer Science, Iowa City, IA, 1999.

20. Kent Lee. A formally verified register allocation framework. *Electr. Notes Theor. Comput. Sci.*, 82(3), 2003.

21. Peter Linz. *An Introduction to Formal Languages and Automata*. Jones and Bartlett, Sudbury, MA 01776, 2006.

22. John McCarthy. *[Photograph]*. Photograph provided courtesy of John McCarthy and used with his permission., 2008.

23. Robin Milner. *[Photograph]*. Photograph provided courtesy of Robin Milner and used with his permission., 2008.

24. Robin Milner, Mads Tofte, Robert Harper, and David Macqueen. *The Definition of Standard ML - Revised*. The MIT Press, May 1997.

25. P.D. Mosses. Unified algebras and action semantics. In *Proceedings of STACS '89*. Springer-Verlag, 1989.

26. P.D. Mosses. *Action Semantics: Cambridge Tracts in Theoretical Computer Science 26*. Cambridge University Press, 1992.

27. P. Ørbæk. Oasis: An optimizing action-based compiler generator. In *Proceedings of the International Conference on Compiler Construction, Volume 786*, Edinburgh, Scotland, 1994. LNCS.

28. Ruby Home Page. About ruby, 2006. [Online; accessed 22-September-2006].

29. R. Sethi. *Programming Languages: Concepts and Constructs*. Addison-Wesley, New York, NY, 1996.

30. K. Slonneger and B.L. Kurtz. *Formal Syntax and Semantics of Programming Languages*. Adisson Wesley Publishing Company, Inc., New York, NY, 1995.

31. Bjarne Stroustrup. A history of c++: 1979–1991. pages 699–769, 1996.

32. Bjarne Stroustrup. *[Photograph]*. Photograph provided courtesy of Bjarne Stroustrup and used with his permission., 2006.

33. Arild Stubhaug. *The Mathematician Sophus Lie*. Springer, Berlin, Germany, 2002.

34. D. Thomas. *Programming Ruby*. The Pragmatic Programmers, LLC, Raleigh, North Carolina, 2005.

35. Dave Thomas. *Programming Ruby: The Pragmatic Programmers' Guide*. The Pragmatic Programmers, LCC, United States, 2005.

36. J. Ullman. *Elements of ML Programming*. Prentice Hall, 1997.

37. D. Watt. *Programming Language Syntax and Semantics*. Prentice-Hall, Inc., Englewoods Cliffs, New Jersey 07632, 1991.

38. Wikipedia. Charles babbage, 2006. [Online; accessed 14-January-2006].

39. Wikipedia. John vincent atanasoff, 2006. [Online; accessed 14-January-2006].

40. Wikipedia. Prolog, 2008. [Oline; accessed 13-February-2008].

Index